WHY THE HYDRANTS FAILED

AND THE FUTURE OF HOUSING IN THE WEST

ROBERT W. SIMPSON

Dedication

This book is dedicated to the generations, families, individual legal citizens, and legal immigrants who built the highly diverse culture that defines the current state of the United States population and electorate. To this end, and for the benefit of future generations, we all must realize that teaching our youth the complex civics required to sustain a nation where true freedom exists requires an education and immigration system that actually teaches how a freedom-based system of government functions.

Here is a famous quote from Ronald Reagan:

"Freedom is never more than one generation away from extinction. We didn't pass it to our children in the bloodstream. It must be fought for, protected, and handed on for them to do the same or one day, we will spend our sunset years telling our children and our children's children what it was once like in the United States, where men were free."

Here is a famous quote from Thomas Jefferson:

"An educated citizenry is a vital requisite for our survival as a free people."

For the generations of free citizens who built this nation, today's education system is clearly inadequate to train our youth and more recent immigrants in what is required to maintain our status as free people.

Acknowledgments

President Trump coined the term "Freedom Cities" and called for a competition regarding the location, design and function of new cities. Politically, he has opened the door for more conservative thought on this subject matter. Our nation currently has a 4.5-to-7-million-unit housing shortage. While boosting the economy and reducing interest rates are helpful measures that could aid our efforts to house the homeless, new Freedom Cities, each comprising its own jurisdiction, should help restore balance to our democracy. Few things in life can provide a greater sense of patriotism than owning a home. Pride in ownership extends to owning a small piece of our great nation. Addressing the current issue of supply and demand within the housing sector of our economy is an important part of President Trump's plan. Homes that are affordable for this and future generations to purchase and own would be a great legacy for any president.

Generally speaking, all cities have one or more industries that comprise a basis for the micro-economy that supports the population and other small businesses within a city. Detroit has a portion of the auto industry. Los Angeles has Hollywood and the entertainment industry. Building a series of relatively small cities on tiny patches of the vast amount of federal land available throughout the West would help alleviate the current housing shortage and provide greater diversity of opinion within our political system. Cities like Sacramento specialize in political influence and require some form of check and balance. A key issue regarding the location of each new freedom city will be creating a micro-economy that properly supports the new population. No form of housing is affordable without the jobs required to support the local economy.

A second inspiring figure for this book is John Westly Powell. Powell was a Civil War Veteran, naturalist, explorer of the West, and the Second head of the United States Geological Survey. In 1879, he published his "Report on the Lands of the Arid Region of the United States." A key finding in his report that Powell was insistent upon was that the water that exists "should not be separated from the land from which it originates." Shortly after his death, the political power that existed created laws and policies that separated the water from the land. The reasoning behind this injustice was actually quite sound. During the period of time that encompassed the world wars, the threat of a Pacific invasion of the West Coast was real. The population centers required to defend the West Coast would need water and food. Most of the major cities along our West Coast would not exist in their present form without the laws and policies that allowed downstream developers to separate the water from the lands of the arid region.

Solutions to the inevitable issues surrounding land, water, population centers, and even political power already exist. Explaining how we execute President Trump's vision in light of Mr. Powell's findings is the basis of this book. Comparing the images below helps us understand that the primary issue relating to the creation of Freedom Cities is the availability of water. President Trump is correct. We have plenty of land with tremendous natural resources that could support large populations. The resource we need to focus on is fresh water.

Also, I would like to acknowledge OpenStreetMap and the USGS for their contributions and for providing the mapping data used to create the images featured in this project. The map data is available under the Open Database License (ODbL).

The Arid Region of the United States:

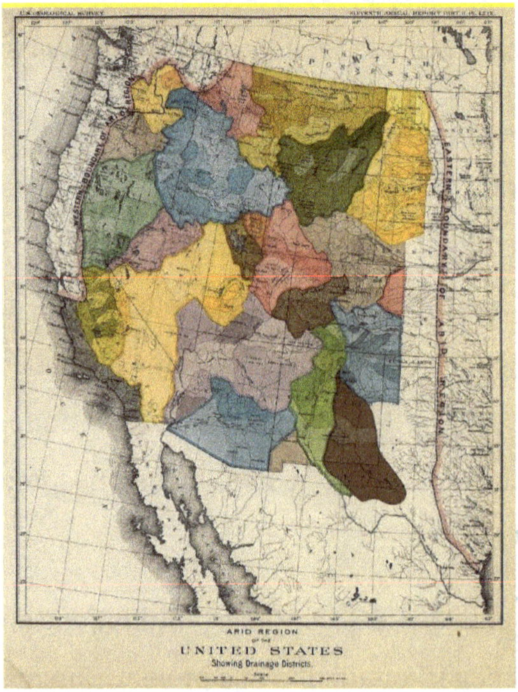

Federal Lands of the United States Map

6 Comments

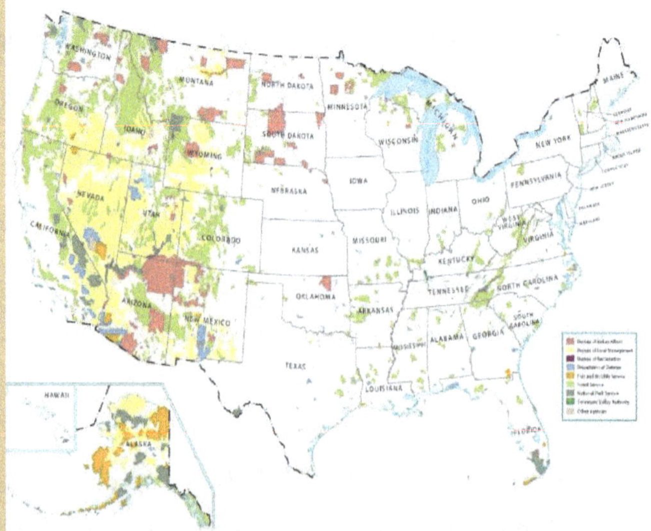

About the Author

Robert W. Simpson was born in Stillwater, Oklahoma, in 1965. His mother had received her BA in home economics, and his father had just completed his dissertation for his PhD in mechanical engineering. Both had attended Oklahoma State University and remained strong supporters of "Pistil Peat" in all his athletic endeavors throughout their lives. In the prior generation, Robert's grandfathers were Oklahoma Rangers and a surveyor from the interstate system. Robert's ancestry on this continent stretches back beyond the formation of our great nation and includes the notable inclusion of the Keith family. Most believe that the Keith family hosted General George Washington and his staff in the days just prior to General Washington's crossing of the Delaware.

Shortly after his birth, Robert was whisked away to sunny Southern California and the City of Placentia in Orange County. Here, his father embarked on a long career designing inertial guidance systems for ICBMs and nuclear-powered submarines. His mother was tasked with raising three sons and being highly involved in the local school board and Baptist church.

Even at this stage, the liberal political leanings of California were well documented but not yet out of control. For his father and all the "redneck rocket scientists" that moved their families to the West Coast, this growing dichotomy of culture would eventually lead to California creating the economic conditions that led Rockwell International and Boing to effectively leave the state. This, in turn, led to what the locals referred to as "the fall of the orange curtain." Orange County was once a conservative stronghold. However, allowing the existence of one or more such strongholds would jeopardize the existence of the "one-party state" we see today. A true diversity of opinions was actively suppressed.

Robert graduated from college at California State University Fullerton. Here, Robert first experienced the full effect of the one-party state, DEI, and the full effect of this particular culture clash. During his sophomore year, Robert had the highest GPA on the starting varsity wrestling team. Robert had fully bonded with a racially diverse team when the athletic director made it clear that he would never receive a scholarship due to his privileged background. While his father made good money for the time, there was no way he could afford to send one or more of his three kids through college and afford to retire at a respectable age (65-70).

From this point, Robert worked his way through his remaining years of college as a framer and then superintendent for a small remodeling and custom home builder. Having an employer that allowed him to Work 32 hours per week during the school year was a true blessing. However, it was also exhausting, and his GPA suffered from it. On the other hand, having taken his pre-engineering courses in calculus and physics allowed Robert to gain a reputation as a "job site MacGyver."

Robert's construction career started with his first real job as a lather in the summer of 1982. A lather installs the paper and wire under the paper and wire under the exterior coat of stucco common in the West. Starting to work at the age of 16 when you got your first car or truck was a cultural value in Orange County at the time. These jobs usually were summer jobs that allowed the local youth to pay for a small portion of their parents' cost of raising their kids. The midwestern phase for this was "pay'n for your raise'n". Today, this cultural value has been replaced by forgiving student loans. How is a parent supposed to raise their kids with proper ethical values when it is the clear intent of the deep state to destroy such ethics? As Robert entered his junior year in college in about 1987, he simply extended his construction career until the housing market crash in 2008. Robert was able to extend his career until 2010 as a development consultant. However, as this form of today's housing crisis developed, the need for a short white Anglo-Saxon male evaporated quickly.

In 2005, a bit of tragedy hit the Simpson family. Robert's father had been diagnosed with Alzheimer's disease in 2003, and his cognitive abilities were fading rapidly. In this same year, Robert had attained full custody of both of his sons and had become aware of the impending housing market crash in 2005. The logical path forward dictated that Robert should sell his home in Semi Valley and move home to provide financial and physical assistance to his mother. After selling his home and moving back into the family home, Robert became responsible for most of the duties involved in running a good-sized family home and managing the decline of his personal wealth. At this time, the cash brought to the table exceeded $120,000. Not much in today's dollars, but it helped the family survive at a critical moment in time. Understand that the landscaping, remodeling, and generalized maintenance provided to the Simpson home was extensive and took years to complete. Over this time frame, the bill for personal labor would have easily exceeded $100,000.

As the market crash of 2008 unfolded, property values plummeted. Debt-to-equity ratios rose dramatically, and lenders were forced to foreclose on far too many questionable loans. This crisis was partially induced by extending home loans to individuals who truly could not afford such loans. In the following years, Robert began to intensely study the issue of how to build truly affordable housing in the arid region of the United States without inducing another market crash. The phrases "affordable housing" and "big city" should never be used in the same sentence as they are antithetical to the existence of each. In fact, they are a contradiction in terms that is exasperated in the presence of high-volume illegal immigration.

The above provides some insight as to the background and raising of the author. Thomas Jefferson and many others believed that the secret to the survival of a free society lies in the education of the electorate. **"An educated citizenry is a vital requisite for our survival as a free people."** At this point in time, what percentage of the American public believes the degrees provided to those who progress through our public education system are worth the paper they are written on?

Table of Contents

Foreword

The intent of this book is to illustrate that the primary obstacle to building new Freedom Cities or even affordable housing on federal lands is freshwater availability for our vast tracts of "underutilized" federal lands. We have plenty of land and natural resources in the arid portion of America's Western frontier. However, most of the water that falls as rain or snow within this region is typically "earmarked" for downstream coastal cities. A portion of the precipitation that falls within the Colorado River basin is also moved East to Denver, Colorado. All of the inland mountain valley lands that comprise the best locations to build new population centers are limited by 100+ year-old legislation and legal precedent dictating that the precipitation that falls on these lands is not available for local use. Just as John Westley Powell feared, it has been monopolized by "downstream developers" that mostly exist in the coastal cities.

For those who witnessed the shocking attempt on President Trump's life during a campaign rally, the concept of divine providence has taken on new significance. As powerful forces within the deep state orchestrated relentless legal and physical attacks against him, we watched his faith and resilience grow stronger each day.

While we mourn the victims of the devastating Los Angeles fires, we must also confront the systemic issues that contributed to this disaster. The utter dependence of coastal cities, depending on precipitation that falls in remote locations as far away as Wyoming is one such issue. It is also important to understand that 80% of the power grid in the US is privately owned, whereas 85% of the California water distribution system is still publicly managed. This discrepancy frequently results in misunderstanding, inefficiency, and blame-shifting when disaster occurs—particularly in California. The established interests managing the state's water distribution system hold significant power, frequently withholding vital information and altering historical accountability. Nonetheless, Americans are becoming more aware of these strategies and has started to completely dismiss them.

Wildfires have always posed a serious threat to communities built upon arid western states. Although enhanced land management and fuel reduction methods are important, they are only half of the solution. The decisions we make about where, what, and how we rebuild will profoundly impact future generations' ability to achieve the American Dream. Even in California, federal lands present an opportunity for innovative and strategic rebuilding efforts.

This book seeks to demonstrate how we can leverage existing technologies to modernize our water infrastructure and construct high-quality "Freedom Cities" on federal lands. This approach may offer the long-sought pathway to preserving the American Dream for ourselves and our posterity. Every parent across this nation should agree: ***our children deserve the opportunity to build a better future.***

Whether the message comes from the muzzle of a gun, a Hurricane impacting North Carolina, or from a wildfire in LA, the good lord appears to be sending us a message. I think the time has come for us to listen and respond.

Please note that President Trump has suggested building 10 new freedom cities on underutilized federal lands in the West. In order to have a substantial impact on our 7-million-unit housing shortage, these cities would need to house a minimum of 500,000 citizens each. Larger cities like this generally don't fit well with the wants and needs of the existing population, the vast stretches of land, water reclamation and reuse concepts. The logistics required to support widespread mining, drilling, ranching, forestry and even data centers may require a more widespread solution. The logistics of developing this region could call for a larger number of interconnected cities. We could look to build 100 new freedom cities with populations of about 50,000 people each. A collection of smaller, interconnected cities spread out over this vast region would likely serve the regional culture, environment, water availability limits, economy and politics better than 10 larger and more widespread cities.

In both examples above, the population increase for the region would be about 5 million people. Our housing shortage is thought to be about 7 million housing units. The average number of residents per household is about 2.6 people per home.

(5,000,000 / 2.6) = 1,923,077 housing units

(1,923,077 / 7,000,000) = 27.47% of our housing shortage

Any form or number of additional housing in the arid West will be limited by physical water availability. Today, 90-95% of indoor water use could be reclaimed and returned to the watershed. We have plenty of land. Water availability and reclamation are the key to increasing populations in our remaining lands of the frontier. Replacing any unreclaimed water that is "earmarked" for coastal cities can easily be replaced with coastal desalination.

Moving forward, I would encourage President Trump, my fellow members of the MAGA movement, and every American citizen to reflect on the wisdom of Matthew 17:20. This is a message of hope!

MATTHEW 17:20

"And Jesus said unto them, because of your unbelief: for verily I say unto you, if ye have faith as a grain of mustard seed, ye shall say unto this mountain. Remove hence to yonder place, and it shall remove, and nothing shall be impossible unto you."

KING JAMES VERSION (KJV)

Chapter 1

Introduction – The California Fires and the Hydrant Failures

The California wildfires are an annual tragedy, but the failure of Los Angeles' water infrastructure and policy during these fires is a man-made disaster, one that was entirely manageable or preventable. As flames consumed homes and businesses, firefighters faced an unthinkable problem: **fire hydrants ran dry.**

In Pacific Palisades, many high-value homes burned to the ground. The issue wasn't due to a shortage of firefighters or equipment but because of poor management of water and forests in California. There were three large water tanks, each holding 1 million gallons, that ran dry after a few hours, leaving firefighters without the water they desperately needed.

A bigger concern was the Santa Ynez Reservoir, which is very large and can store 117 million gallons of water[1]. It was shut down more than 13 years ago. If it had been maintained and filled, it could have been crucial in fighting the fires. Why was this reservoir, meant for emergencies like these kept empty for so long?

This situation highlights how government errors, bad management, and political issues have complicated what should have been a straightforward emergency response.

---

[1] Ingram, J., & Sarwate, J. (2025, January 14). U.S. When L.A. fires broke out, the 117-million gallon Santa Ynez Reservoir near Pacific Palisades was empty. Here's what we know. CBS News. https://www.cbsnews.com/news/la-fires-santa-ynez-reservoir-pacific-palisades-california/

The Broken Water System: Who is Accountable?

California's highly centralized water system is in the center of this crisis. It is overseen by state-controlled monopolies like the Los Angeles Department of Water and Power (LADWP) and the Metropolitan Water District (MWD). Unlike private utilities, which face strict regulations, competition, and potential lawsuits for negligence, these government-run agencies function with almost complete immunity.

The image below illustrates that the portion of publicly owned water distribution systems is about 52%. In California, the percentage of the water system owned by the state jumps to 90%.

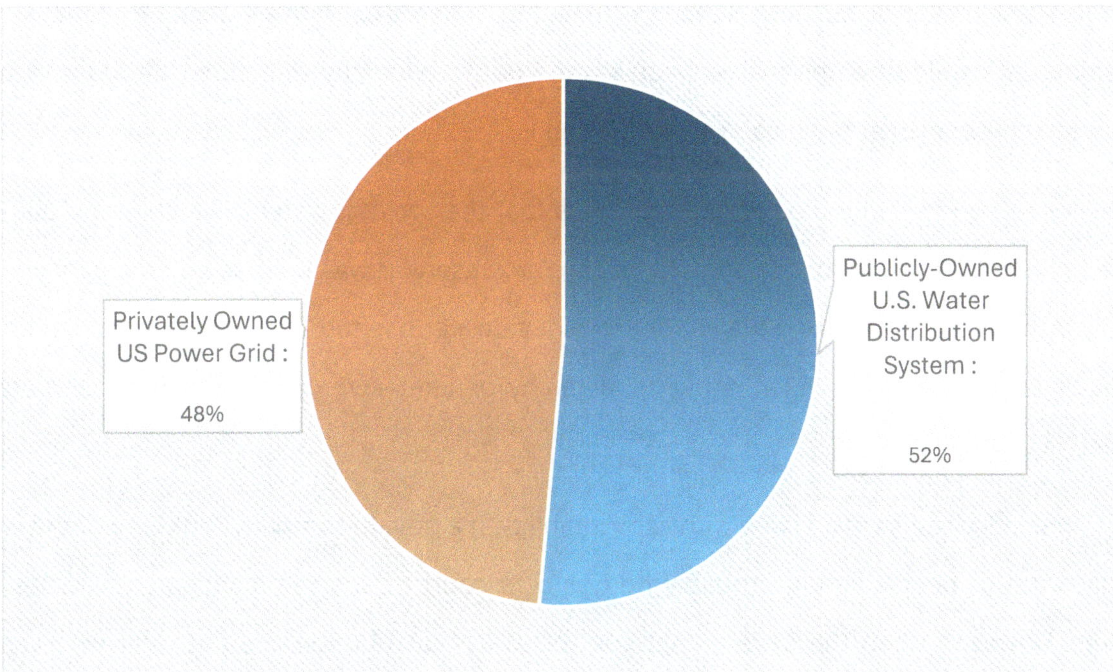

Over 90 percent[2] of the state's population receives water through approximately 400 large community water systems, each of which operates 3,300 or more service connections. This creates a massive and unresponsive bureaucracy. These systems are typically urban retail utilities that serve a minimum of 3,000 homes and businesses each. The secret to the political power and regulation behind the California water distribution system is understanding who owns the canals and aqueducts. For instance, the California Aqueduct is owned by the State of California. The

[2] Drinking Water Sources: An Overview. (2024, January 10). CDC. https://www.cdc.gov/drinking-water/about/drinking-water-sources-an-overview

Mokelumne aqueduct is owned by the East Bay Municipal Utility District (EDMUD). The Hetch Hetchy aqueduct is owned by the San Francisco Public Utilities Commission. The Los Angeles aqueduct is owned by the City of Los Angeles and the LADWP. While some publicly traded "public utilities" exist, this thin veil of participation in our capitalist system is as transparent as Chinese (CCP) ownership of firms traded on the New York Stock Exchange. China "owns" a majority interest in at least 2,400 companies traded in our "free market" stock exchanges. Furthermore, an additional 263 companies "born and bred" in China are traded in our markets. The Socialist "one party state" of California that controls the vast majority of the water distribution system throughout the arid west and uses this control to subsidize development where and when they want it. The concept of building "Freedom Cities" outside of California's ability to control such development would threaten the well-established public oligarchy that currently has a massive ability to achieve market price control on housing.

A major reason for "why the hydrants failed" is that the socialist water oligarchies in California placed revenues above public safety. The Santa Ynez reservoir was built as an emergency water supply. After it was built, California sought to reduce water use through conservation. An empty reservoir evaporates zero water. A public oligarchy generates zero revenues from evaporated water.

This imbalance results in an issue with accountability. If a private company took care of the water supply, a lot of people would probably be going to court to sue the managers for issues and failures. However, when the government controls water distribution, it's not clear who can be blamed or held responsible for the issues. This lack of clarity makes it difficult to address problems and ensures accountability.

The Santa Ynez Reservoir Scandal

The Santa Ynez Reservoir has been an essential part of Los Angeles' fire safety plan since it was built in 1972 to accommodate the growth in Pacific Palisades and enhance fire protection. This reservoir serves as a backup water source, ensuring robust firefighting capabilities during high-wind firestorms. Its significance was underscored during the 2025 Palisades Fire, highlighting its essential role in safeguarding the community against the ever-present threat of wildfires. Yet 14 years ago, city and state officials deliberately took it out of service with no clear explanation.

- Was the reservoir leaking?
- Were structural issues discovered?
- Was the public informed of the consequences?

The answers to these questions remain elusive. Despite advances in technology, there appears to be no rational reason for keeping this massive reservoir dry. Some speculate that the move was politically motivated, part of a larger push to centralize water management under state control. Others speculate that leaving many of the local reservoirs in Southern California dry would reduce evaporation rates from filled reservoirs across the system.

The Lack of Resource Management and Accountability

During the 2025 wildfires in Los Angeles, the water supply system faced significant strain. The high demand for water during firefighting efforts led to low water pressure in elevated areas, causing fire hydrants to run dry. The Los Angeles Department of Water and Power reported that water tanks in the Palisades were depleted due to four times the normal demand for 15 hours straight. At the 27:30 time stamp, the Chief Engineer for the LADWP indicated that the three 1,000,000-gallon water tanks associated with the local portion of the distribution system ran dry, and the fire hydrants in the upper regions of Pacific Palisades also went dry.

A critical issue with the explanation provided by LADWP engineer Quinones is that the system was designed to provide treated drinking water to local "taps" and fire hydrants and to supply sufficient water to fight fires in non-firestorm regions. However, this region is highly prone to Santa Ana windstorms, which concentrate wind energy through the local hills and passes.

In short, the homes and other structures in Pacific Palisades exist in a natural "blast furnace," and the water system was not designed to provide enough water to protect such structures from a firestorm unless the Santa Ynez Reservoir is full at the time of the firestorm.

The combined 3 million gallons of water storage plus the engineered water flows into the local area were clearly insufficient to fight the fire. This raises the question: why has the Santa Ynez Reservoir been "out of service" for the last 13 years? It appears that the flows from the pipeline, combined with the 3,000,000 gallons in the storage tanks, were far short of the water required to fight any fire in this canyon during a Santa Ana wind event.

Reflecting on the history of Pacific Palisades, it is extremely probable that the developer paid for and built the storage tanks and the Ynes Reservoir as a "condition of approval" to meet the fire codes of the time. The threat was well-established and widely known. However, about 14 years ago, those in charge made a conscious decision to take the Ynes Reservoir out of use, making it "unavailable."

All water passing through the water conveyance system that enters the area of the Pacific Palisades fires came from the combined flows of the California Aqueduct and the Los Angeles Aqueduct. In the early hours of fighting the fire, every drop of the water that fed the fire hydrants and 3,000,000 gallons of water tank storage was treated in the Jensen Plant. All of this water is acceptable for use as tap water.

The line drawings for the exact route of the pipeline that feeds Pacific Palisades can be found as public records. However, obtaining them would likely require an approved FOIA request and a lawsuit. This version of the existing pipeline measures 20.6 miles from the Jensen Plant to the Santa Ynez Reservoir. Please note that the number of "laterals" coming off of the line shown is large. Everyone needs water.

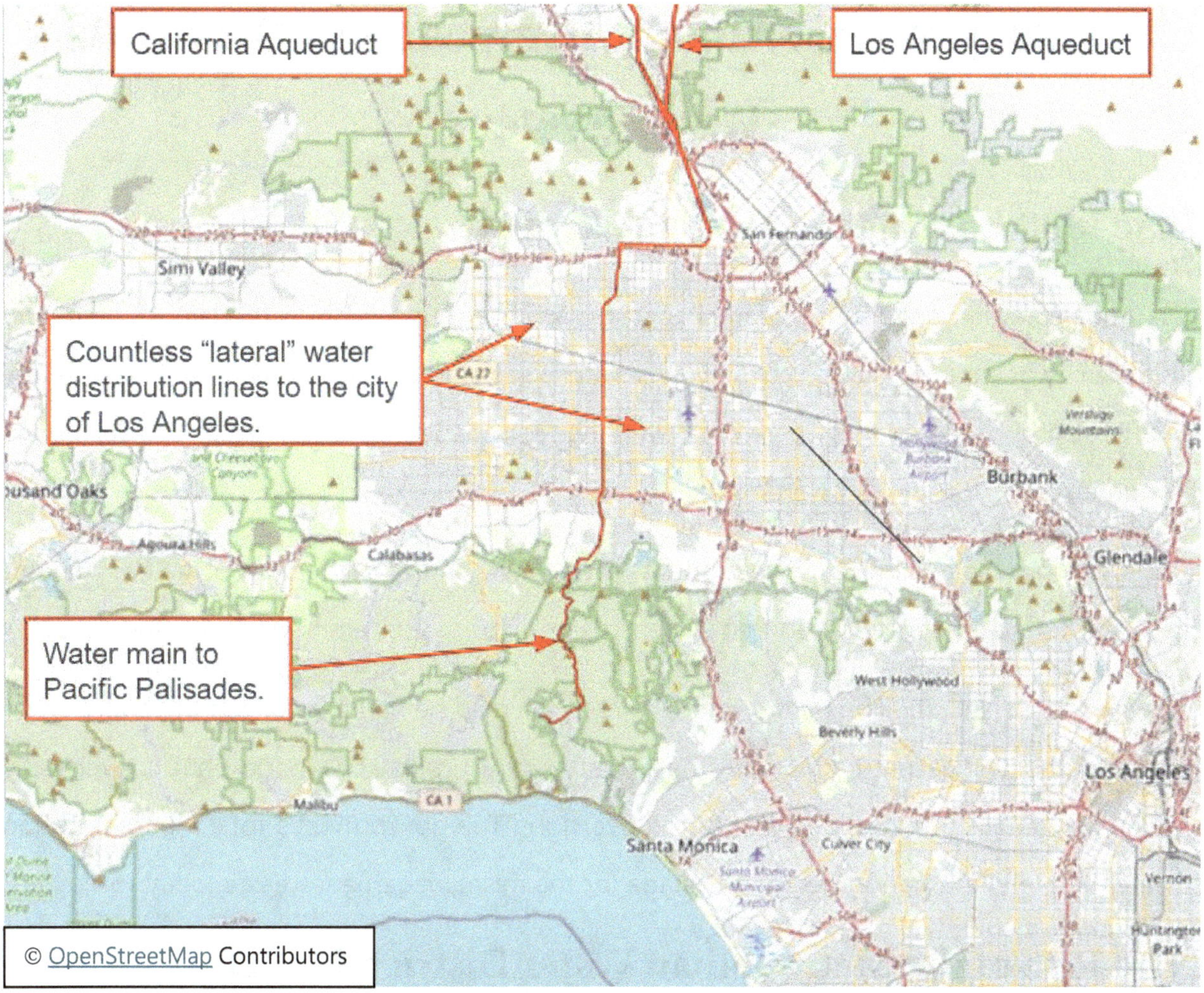

(Pacific Palisades)

It is clear that the combined total of water available to fight the fires in Pacific Palisades was insufficient for the protection of life and property. Local water availability to the fire and garden hoses in Pacific Palisades can be generalized as the water flowing through the pipeline plus the 3 million gallons stored in the local tanks.

Using visual satellite imagery, it is easy to find the first two out of the three 1,000,000-acre-foot storage tanks referred to in the news conference. The third tank may be buried underground, usually referred to as a cistern. This Satna Ynez reservoir holds 118,000,000-gallons and appears to have been removed from service in or about 2010.

A History of Government Overreach in California's Water System

The 1898 Los Angeles Water Takeover

Los Angeles' water system has been shaped by political interests since the late 19th century when the city government seized control of its water supply from private hands. Fred Eaton, a key city official, proposed a system where water would be provided to residents free of charge, funded through taxes.

But even from the beginning, corruption was evident. When the city took over the Los Angeles City Water Company, it discovered that no records or maps of the water system existed.

William Mulholland, the company's chief engineer, claimed he had memorized the entire water system. Mulholland secured a job with the city after proving his knowledge—but conveniently omitted key details that would have ensured oversight. The government's total monopolization of water began in this period, setting the stage for the mismanagement we see today.

The Rise of the Metropolitan Water District (MWD)

Founded in 1928, the Metropolitan Water District (MWD) was created to manage large-scale water projects, such as the Colorado River Aqueduct and the California Aqueduct. Today, it serves over 19 million people[3] across Southern California and controls a staggering 5,200 square miles of land.

While MWD provides the necessary infrastructure, it also represents an unchecked socialist-style monopoly where water supply decisions are made by unelected officials. With no market competition and no real accountability, inefficiency and mismanagement are inevitable.

---

[3] Metropolitan to Further Reduce Reliance on Colorado River with $186 Million Federal Investment. (2025, January 10). The Metropolitan Water District of Southern California. https://www.mwdh2o.com/press-releases/metropolitan-to-further-reduce-reliance-on-colorado-river-with-186-million-federal-investment/

The Pacific Palisades Fire: A Case Study in Government Failure

The 2024 LA wildfires exposed the full extent of the water system's dysfunction. Firefighters on the front lines were shocked to find that many fire hydrants had lost pressure or were completely dry.

The official explanation from LADWP was that the three 1,000,000-gallon water tanks serving the area had been drained after 15 hours of battling the fire. Wildfires can last for days or weeks. But the reality is far worse:

- The Santa Ynez Reservoir, with 118 million gallons of capacity, was unavailable or unutilized.
- The 20.6-mile pipeline from the Jensen Water Treatment Plant could not supply enough water.
- Firefighters were left with severely limited water access in one of the most fire-prone regions of California.

Had the reservoir been operational, thousands of homes could have been saved. However, due to bureaucratic neglect, an essential safeguard against wildfires had been deliberately dismantled.

Urban Sprawl and the Water Supply Crisis

California's urban sprawl presents another challenge. Unlike compact cities like New York or Boston, Los Angeles is spread out over thousands of square miles.

Comparison of Population Density (People per Square Mile):

- Manhattan – 70,826[4]
- New York City – 29,302[5]
- Boston – 13,841[6]
- Los Angeles (City) – 8,304[7]

---

[4] Manhattan Population 2022. (n.d.). Azurewebsites. https://poeent.azurewebsites.net/manhattan
[5] United States Population Density. (n.d.). Kontur Inc. https://www.kontur.io/datasets/population-dataset/united-states-population-density/
[6] McAllister, R. (2024, September 25). Boston Population 2025 – Key Statistics and Trends. North American Community Hub Statistics. https://nchstats.com/boston-population/#google_vignette
[7] QuickFacts Los Angeles City, California. (n.d.). Census. https://www.census.gov/quickfacts/fact/table/losangelescitycalifornia/PST045224

- Los Angeles (County) – 2,344[8]

This means that Los Angeles requires hundreds of miles of additional water pipes, creating enormous strain on the system. When thousands of residents turn on their hoses during a fire, water pressure drops, fire hydrants lose effectiveness, and entire neighborhoods become vulnerable.

California's poorly planned water infrastructure cannot support this level of urban sprawl.

A Smarter Path Forward

Instead of rebuilding within a broken system, California should embrace new solutions—such as President Trump's vision for Freedom Cities.

What Are Freedom Cities?

Freedom Cities are privately developed, high-tech communities built on federal land—outside the constraints of California's bureaucratic control.

Instead of relying on state-controlled utilities, these cities would:

- Use privately managed water systems with redundant reservoirs to prevent shortages.
- Incorporate smart infrastructure that dynamically adjusts water distribution during emergencies.
- Implement fire-resistant building materials and zoning laws that prevent high-risk developments.

If California's state government cannot manage basic infrastructure, it is time to bypass the bureaucracy and build new communities that can thrive independently.

[8] Demographics of Los Angeles County. (n.d.). Retrieved April 23, 2025, from https://en.wikipedia.org/wiki/Demographics_of_Los_Angeles_County

A System in Need of Reform

The LA fires were not merely a natural disaster; they were the predictable result of government incompetence. A deliberate decision to take a critical fire reservoir out of service left communities defenseless.

As Einstein famously said:

> *"Insanity is doing the same thing over and over again and expecting different results."*

California's state-controlled water monopolies have failed. If we do not change course, these failures will repeat year after year. The time for radical reform is now.

Whether through privatization, legal action, or the development of Freedom Cities, America must take back control of its infrastructure before another preventable tragedy strikes.

Chapter 2

A History of Over Development and Water Issues in the West

The history of water use and distribution in the United States, particularly in the arid West, is intrinsically tied to the broader narrative of the nation's expansion and development. Since the Lewis and Clark expedition in the early 1800s, the issue of water rights has played a key role in shaping both politics and the economy. Today, this is still true as seen in the ongoing water conflicts in California. In the Western United States, the way people have come to understand and manage water resources has been both a blessing and a problem. Managing water effectively has allowed people to transform harsh, dry lands into areas where growing cities can be built. However, it has also presented many challenges along the way.

Since the beginning of the state's history, managing water resources has been a significant concern. The demands of agriculture, a quickly expanding population, and the dry weather are the causes of this problem. Large tracts of farmland were converted to cities as Los Angeles flourished. Orange groves, for example, used to cover Orange County, but they are no longer there. Large aqueducts, dams, and reservoirs were constructed to meet the growing demand for water, drastically altering California's landscape.

While these engineering feats secured water for cities, they also introduced lasting challenges, including the controversial practice of water reclamation, often referred to as "toilet to tap." Building such systems close to the beach means that manmade devices must accomplish the entire process. Reclamation facilities that discharge high-grade effluent water into aquifers or riparian systems and allow Mother Nature to finish the process are the preferred method. Unlike irrigation, where much of the water is lost to evapotranspiration and contributes to downwind precipitation, reclaimed water is cycled directly back into municipal supplies, a reality many citizens remain uneasy about.

As the coastal cities grew, the length of underground water pipes grew at an exponential rate. If such a city had a circular footprint, the length of underground water pipes would grow at a rate proportional to r^2 in the formula:

$$A = \pi r^2$$

Maintaining such a distribution system would become expensive. We would have to make sure that local water reservoirs are provided, maintained, and filled. This would help manage emergency water needs.

The 20th century saw the most significant of these efforts, with projects like the Los Angeles Aqueduct, the Colorado River Aqueduct, and the California State Water Project becoming central to the state's survival and prosperity. The purpose of these extensive water diversion networks was to supply the demands of a growing metropolitan population. However, some things turned out differently than expected. As the population grew, so did the demand for water. This growth put more strain on already overworked water supplies. William Mulholland and various engineers played a key role in addressing these challenges. They initiated extensive projects that ultimately evolved into the complex systems seen today, ensuring that millions in the state have access to water.

The Era of Manifest Destiny and Early Expansion

Manifest Destiny was a powerful belief among Americans during the early nineteenth century. They believed that the United States was destined to expand throughout North America. That certainly involved the Lewis and Clark Expedition, which occurred from 1804 to 1806. Discovering the land, riches, and opportunities in the West, they mapped a path to the Pacific Ocean. The new land was well-sought by many entrepreneurs and inhabitants.

Individuals discovered it simpler to head west during the 1810s due to Robert Stuart's forays along the Oregon Trail during 1812 and 1813. This led to an era of active land exploration and settling. Yet, when migrating to the west, they were facing a hostile and challenging landscape. The country was not readily compliant with the visions of Manifest Destiny. It then became apparent that infrastructure, such as buildings and roads, was required for these new communities, particularly in areas that were dry and hard to develop.

The Trail of Tears: Forced Displacement of Native Americans

Between 1830 and 1850, the US government forcibly moved Native American tribes to clear land for European settlers. This tragic occurrence, known as the Trail of Tears, forced thousands of Native Americans to abandon their homes in the southeastern U.S. and elsewhere. It aided in moving the nation to settle more toward the West. The enforced displacement revealed a conflict between the aspiration of Manifest Destiny, where Americans thought they should possess all the territory from coast to coast, and the reality that numerous Native American tribes had inhabited these areas for generations.

As the nation expanded, another major problem came to light. Individuals had to determine ways to distribute water. Water was essential for agriculture, constructing shelters, and sustaining communities, particularly in the arid areas of the West.

John Wesley Powell and the United States Geological Survey (USGS)

As the government of the US extended westward, land and water management became a pressing concern. John Wesley Powell, an explorer and geologist, was a key person in addressing these concerns in the mid-1800s. He assessed the dry regions of the West. Before this, he worked as a mapmaker and land surveyor for the Union Army during the Civil War. In 1881, he became the second director of the United States Geological Survey (USGS).

Powell's experience with the land and resources of the West led him to see that land and water in dry areas needed careful management for people to live a better life. In 1879, he wrote a report for Congress called "Lands of the Arid Regions of the United States." This report stressed that water rights should remain linked to the lands they water. He explained that separating land and water could lead to unsustainable growth and harm the environment.

Powell's report discussed the complex relationship between water and land in the western United States. He recommended setting up a federal system to manage these water resources efficiently. However, many people ignored his advice because his ideas were at odds with the expanding commercial interests. Business people and developers, in particular, were keen to control water access to make money. They aimed to dominate water rights to expand their land ownership and increase their profits. Powell's ideas were visionary, and while they did not immediately influence policy, they foreshadowed the battles that would come later in the 20th century, which we now recognize as the "Water Wars" of the American West.

From Powell's perspective, no water should have been allowed to be exported from the arid region to coastal cities. Today, building affordable housing on federal lands has bipartisan support. President Trump called for a contest to find the best concepts and locations to build new freedom cities on federal lands.

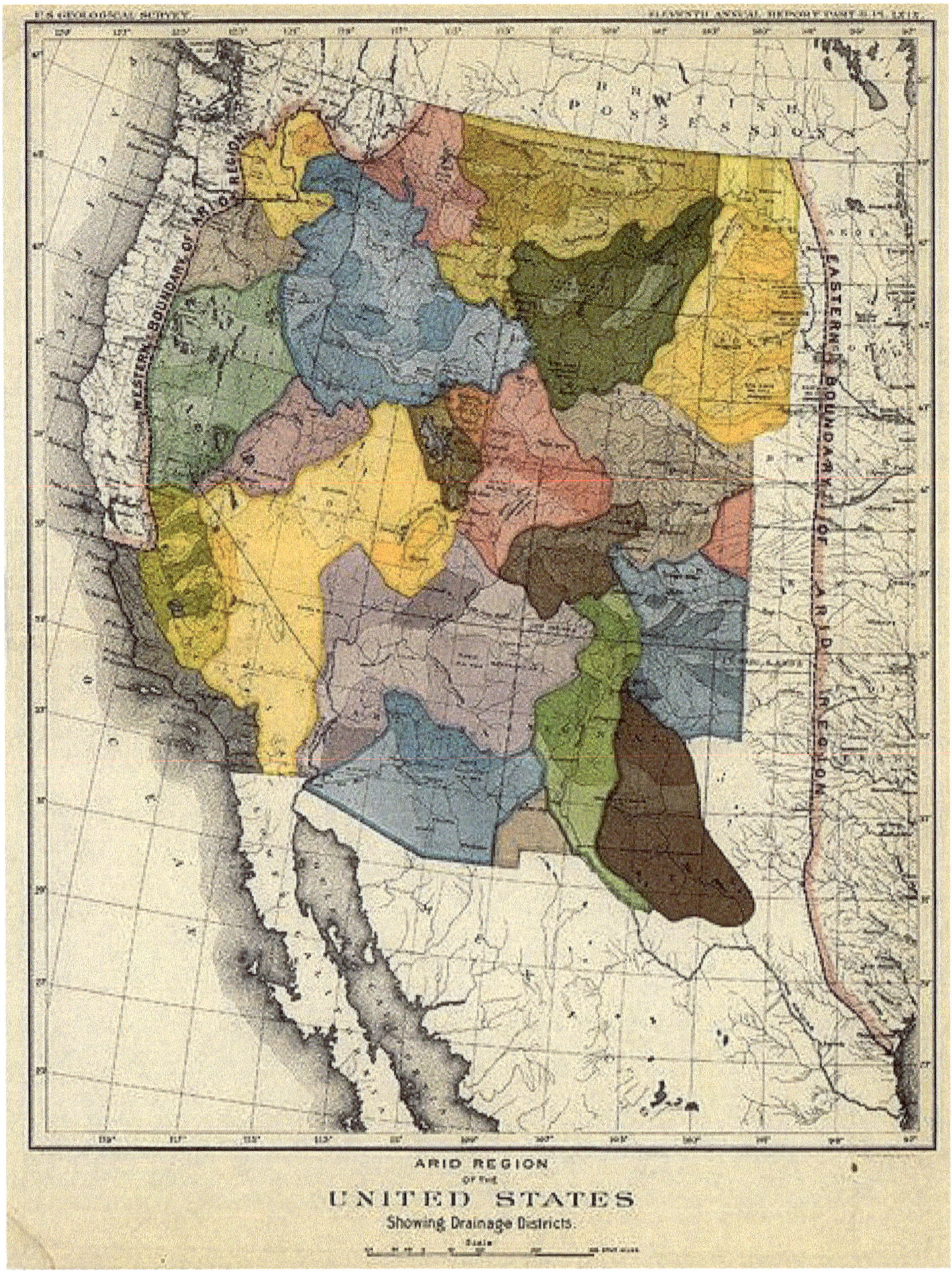

A primary reason why the remaining federal lands in the US overlap so well with "The Lands of the Arid Region of the US" is that coastal cities found a legal loophole called Pueblo water rights, and this legal concept was codified by the ratification of the Colorado River Compact.

Each new freedom city built on the arid lands of the US will need its own water supply and reclamation system. In order to follow the principles set forth by DOGE, the city location criteria should conform to the following: Bring the cities to the water, not the water to the cities.

Federal Lands of the United States Map

6 Comments

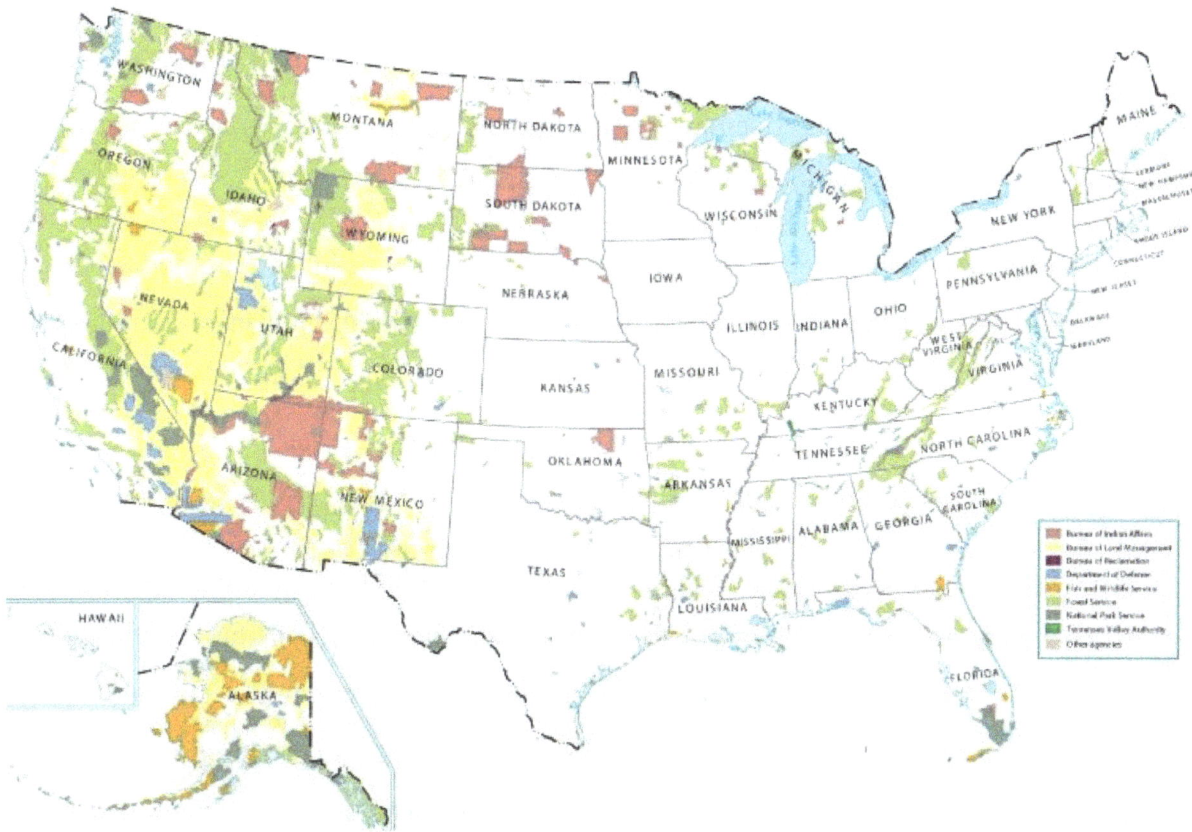

Water Wars and Early Infrastructure

The early 20th century saw the growing dominance of urban areas, particularly along the West Coast, where cities like Los Angeles, San Francisco, and San Diego were rapidly expanding. The need for water in these cities led to a series of ambitious engineering projects, each of which would leave a lasting legacy in the region. This led to the establishment of water rights laws, which were often at odds with Powell's insistence on keeping water tied to land.

Landowners' ability to utilize or redirect water from natural sources on their property was greatly limited in the dry Western regions. Water from rivers and streams was considered public property according to the "riparian rights" doctrine. So, landowners couldn't claim exclusive rights to it without getting particular permits.

The swift urban growth indicated that cities such as Los Angeles, San Francisco, and San Diego required more water to accommodate their increasing populations. This often involved redirecting water from different regions. The advancement of major water infrastructure initiatives highlighted the growing disparity between regions with abundant water resources and those struggling to provide water to increasingly populated areas.

The Los Angeles Aqueduct was among the most significant of these initiatives. The chief engineer of the Los Angeles Bureau of Aqueducts, William Mulholland, managed the building of the aqueduct. The structure was completed between 1908 and 1913, and it was designed to carry water from Owens Valley to Los Angeles. The initial victim of the rising need for urban water sources was Owens Valley, a fruitful area that had once been seen as an important agricultural hub. The building of the aqueduct sparked significant debate as it drained the valley's water resources, leaving local farmers without water and leading to hostility.

This building of this aqueduct had significant consequences on politics as it facilitated the onset of the "Water Wars" involving Los Angeles and various other areas in California. In 1924, communities in Owens Valley and nearby regions fought to regain control of the water being redirected to Los Angeles, signaling the beginning of numerous legal and political disputes regarding water rights. The history of water in the American West is largely characterized by strife and contention, as the needs of rural communities clash with those of urban regions.

Despite the objections, Los Angeles was able to continue grow rapidly due to its ability to secure water supplies. It was the fifth-largest city in the United States by 1930, with a population of over 1.2 million people. The completion of the first Los Angeles Aqueduct in 1913 and the construction of the second Los Angeles Aqueduct between 1965 and 1970 helped fuel the city's growth, providing a steady supply of water that allowed agriculture, industry, and housing developments to flourish. Currently, the Los Angeles Aqueduct is capable of delivering approximately 480 cfs (300,000 acre-feet per year)[9]. In a normal year, it can provide 2.4 million acre-feet per year of water.

As the coastal cities grew, so did their political and legal power. Perhaps the ability of these cities to win lawsuits was the first large and ongoing example of "lawfare" in US history.

The Development of the Colorado River Aqueduct

As Los Angeles expanded, so did its demand for water. In the 1930s, securing enough water from the Colorado River became a major concern, as this river was a crucial water source. To address this, the Colorado River Aqueduct was constructed between 1933 and 1935. This vital aqueduct transports water from the Colorado River at Lake Havasu to Southern California. The construction of the aqueduct was an enormous engineering feat, reflecting the creativity and determination of that era. It also represented an important moment in the history of the American West regarding water rights.

The project, created during the Great Depression, hired hundreds of individuals and became synonymous with both controversy and hope. A water dispute that would be the base of the state's water politics began when water from the Colorado River, which was shared by several states, was diverted to Los Angeles and the surrounding communities. While a legal basis for the allocation of the river's resources was developed in the Colorado River Compact, adopted in 1922,

[9] Owens Valley Water History (Chronology). (2008, January). County of Inyo California. https://www.inyowater.org/documents/reports/owens-valley-water-history-chronology/

conflicts continued, particularly as Los Angeles expanded. The building of massive reservoirs, pumping stations, and aqueducts eventually became a necessary part of the state's infrastructure.

The Colorado River Aqueduct was an element of a broader strategy for managing and allocating water resources. During this period, initiatives such as the Hoover Dam, constructed between 1931 and 1936, were also undertaken. These initiatives sought to control the mighty Colorado River, supplying crucial water storage and electricity to surrounding areas. Although praised for their engineering successes, these projects posed difficulties. They affected the Native American tribes residing close to the river and disturbed the natural habitats and ecosystems as a result of building dams and canals.

The legal basis for distributing water from the Colorado River was established in 1922 when the Colorado River Compact was ratified, commonly known as "The Law of the River." The river was split into an Upper and Lower Basin under the compact, and each region received a certain amount of water. However, the compact was based on a flawed assumption about the amount of water the river could provide. As the demands of urban centers like Los Angeles and Phoenix grew, tensions began to mount over the distribution of water.

A rough comparison of the pre and post-development flows out to sea may provide some clarity on the scale of environmental damage that has occurred over the last 100 years. Prior to development, freshwater flows out of the Bay Delta and, after the conjunction of the Sacramento and San Joaquin rivers, was about 30 million acre-feet per year. Today, it is closer to 40 million acre-feet per year[10]. The predevelopment flows out of the sea of the Colorado River were about 12.5 million acre-feet per year. Today, only 50,000 (0.05 million) acre-feet[11] of reclaimed water flows through the Colorado River estuary into the Gulf of California. Is this a good example of our modern sense of environmental justice?

---

[10] Ring, E. (2025, April 30). Ringside: How to Add 10 Million Acre Feet Per Year to California's Water Supply. California Globe. https://californiaglobe.com/fr/ringside-how-to-add-10-million-acre-feet-per-year-to-californias-water-supply/

[11] Colorado River Water: Good to the Last Acre-Foot. (n.d.). SIAM. https://www.siam.org/media/ljkpwaoe/m3challenge_colorado_river_water.pdf

Large-Scale Water Diversions

The Mokelumne and Hetch Hetchy aqueducts provided similar solutions in Northern California. For the East Bay Municipal Utility District (EBMUD), the Mokelumne Aqueduct brought water from the Sierra Nevada's Mokelumne River. According to the current water rights agreements, the EBMUD can withdraw up to 364,000 acre-feet (0.449 km3) per year[12] from the Mokelumne River.

Meanwhile, the Hetch Hetchy system, constructed in the 1920s, brought water to San Francisco. The controversy surrounding the Hetch Hetchy project was particularly heated, as environmentalists fought to preserve the valley, which was once a pristine wilderness, while others argued that the benefits of the water supply outweighed the loss of such natural beauty. The completion of O'Shaughnessy Dam in 1934 flooded the Hetch Hetchy Valley, creating a reservoir that continues to supply about 265,000 acre-feet of water to 2.6 million people[13] in 29 Bay Area today.

The Mokelumne Aqueduct, spanning approximately 94.5 miles, is a vital conduit supplying water to 35 municipalities in the East Bay, serving around 1.4 million residents[14]. Its diameter varies between 5 feet 5 inches and 7 feet 3 inches, with the Mokelumne #3 Aqueduct consisting of a 10-mile, 7.5-foot diameter steel pipeline. Supplementing its primary supply from the Mokelumne River, EBMUD can also draw up to 98 million gallons per day from the Sacramento River, though this reserve is typically utilized only in the driest 35% of years. Meanwhile, the Hetch Hetchy system, constructed in the 1920s, brought water to San Francisco. The controversy surrounding the Hetch Hetchy project was particularly heated, as environmentalists fought to preserve the valley, which was once a pristine wilderness, while others argued that the benefits of the water supply outweighed the loss of such natural beauty. The completion of O'Shaughnessy Dam in

[12] Mokelumne Aqueduct. (n.d.-a). Retrieved May 4, 2025, from https://en.wikipedia.org/wiki/Mokelumne_Aqueduct

[13] Hetch Hetchy Water and Power System. (2024, August 14). Maven's Notebook. https://mavensnotebook.com/explainers/where-does-californias-water-come-from/hetch-hetchy-aqueduct/

[14] Mokelumne Aqueduct. (n.d.-a). Retrieved May 4, 2025, from https://en.wikipedia.org/wiki/Mokelumne_Aqueduct

1934 flooded the Hetch Hetchy Valley, creating a reservoir that continues to supply about 265,000 acre-feet of water to 2.6 million people in 29 Bay Area today[15].

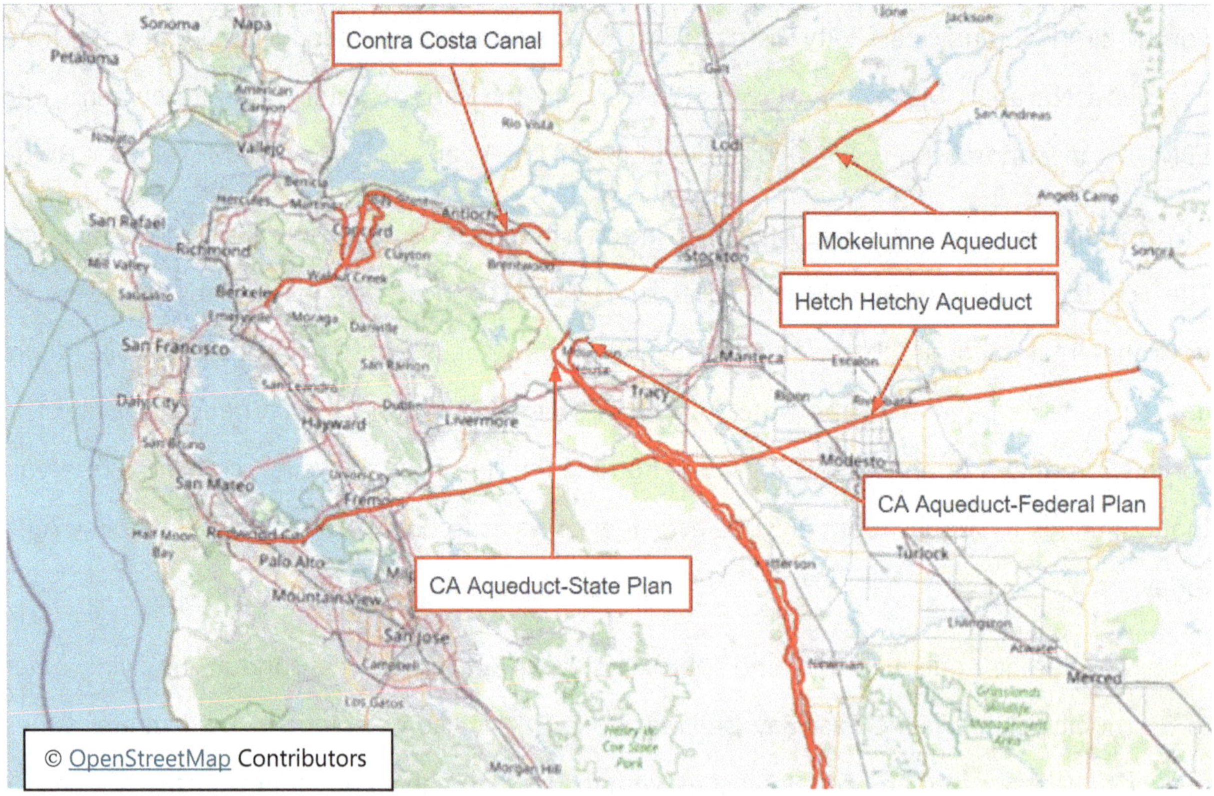

(Bay Area Aqueducts and Canals)

The Hetch Hetchy Aqueduct stretches 167 miles across the Central Valley. Before reaching the Bay Area, it runs through the Irvington Tunnel near Fremont, then divides into four major pipelines—Bay Division Pipelines (BDPL) 1, 2, 3, and 4—each ranging from 60 to 96 inches in diameter. These pipelines cross the Hayward Fault and the San Francisco Bay, with the water ultimately stored in local reservoirs, including Calaveras, Crystal Springs, and San Antonio.

As the state's population exploded, California's reliance on imported water intensified. Supplying water to the growing population in cities is a big challenge because it needs a lot of energy to move water over long distances. The Edmonston Pumping Plant plays a vital role in getting water to Southern California. It's located at the northern edge of the Tehachapi Mountains. This plant

---

[15] Hetch Hetchy Water and Power System. (2024, August 14). Maven's Notebook. https://mavensnotebook.com/explainers/where-does-californias-water-come-from/hetch-hetchy-aqueduct/

supplies approximately 2.5 million acre-feet of water to 13 Southern California water contractors and is the tallest single-lift pumping plant in the world.

The facility employs fourteen large pumps that take water almost 2,000 feet above the Tehachapi Mountains. This great engineering splits water into two broad routes to provide it to all of Southern California. The initial route, the West Branch Aqueduct, transfers water to Los Angeles and surrounding cities, stockpiling it in Pyramid Lake and Castaic Lake. The second route, the East Branch Aqueduct, supplies water to the Inland Empire, extending to areas such as San Bernardino and Riverside, with storage in Silverwood Lake and Lake Perris.

The Edmonston Pumping Plant consumes a tremendous amount of energy as it uses 40% of all the electricity produced for the State Water Project (SWP). It is the largest electricity consumer in the project. The pumps and the plant as a whole are essential for a reliable water supply to Southern California. Yet, they demonstrate the hard challenge of balancing water supply with energy consumption and sustainability objectives.

The reliability and longevity of California's water infrastructure rely in large part on the Edmonston Pumping Plant. The California Department of Water Resources warns that a prolonged interruption of the State Aqueduct, especially between the Bay Delta and the Edmonston Pumping Plant, would have severe consequences. A four-and-a-half-month hiatus would inflict $1 billion in damage and disrupt water supplies to a large percentage of the state's population. Maintaining the equilibrium among urban needs, agriculture, and the environment has become increasingly hard due to these water systems' unrelenting growth. Along with draining rivers and lakes, the diversion of water has altered whole ecosystems.

Mono Lake draining, a leading focus of the debate surrounding the environmental expenses of Los Angeles' water sources, was another of the clearest examples. Los Angeles continued to gain increased scrutiny for the impact it had on the environment in the following decades after it constructed its aqueducts. The increasing clash between the requirements of city populations and the defense of natural resources was highlighted through attempts to renew ecosystems affected by water diversions, either through legal action or public protest.

The State Water Project and Future Strain

The complicated nature of overseeing water resources throughout California prompted the development of numerous supplementary infrastructure initiatives, like the California State Water Project, which seeks to transport water from the state's northern to southern regions. The Governor Edmund G. Brown California Aqueduct, the project's focal point, provides water to millions of people and is essential to Southern California's cities and Central Valley's agricultural. However, the strain on the system increased along with the boom in water consumption, which was caused by a growing population.

Farmland between Los Angeles and San Diego has all but vanished as residential usage has supplanted agriculture as the primary use of the water supplied to the Los Angeles region. Both corporate and family farms in the southern Central Valley are failing as a result of strong political pressure to divert an even larger portion of the water from the California Aqueduct to metropolitan areas. Household water needs take precedence over agricultural supply, which results in lower farming output and higher food prices, further taxing the economy.

Concerns about water scarcity are made worse by the state's current housing crisis. Reliable, sustainable water management is more important than ever since an estimated 1.5 million to 7 million new houses are needed to meet demand.

The Role of Government Agencies and Socialism in Water Distribution

The increasing water needs of the West in the 20th century led to the establishment of many organizations which were responsible for regulating and overseeing water use. Founded in 1902, the Los Angeles Department of Water and Power (LADWP) evolved into one of the most influential entities in the region to oversee the city's water resources. The operations of LADWP grew alongside the population of the city of Los Angeles. By the 2000s, the department supplied approximately 197 billion gallons of water annually to 676,000 residential and commercial clients, ensuring the city's vast communities and industrial centers had the necessary resources to thrive.

Similarly, the Metropolitan Water District of Southern California (MWD), established in 1928, became a key player in the region's water management. The MWD was responsible for overseeing the distribution of water to key cities in Southern California, including Los Angeles, San Diego, and Riverside. It would source the water from the Colorado River and additional locations. The MWD would oversee the construction of additional pipelines and aqueducts over the years, enhancing the region's capacity to transport water from distant places. Throughout the coastline, the California Coastal Commission consistently denied and closed down nuclear power plants and desalination facilities.

To deliver water from the mountain and high desert areas to the coastal cities, Southern California's water system grew into a huge system of dams, aqueducts, and pumping plants by the mid-20th century. Construction of reservoirs like the San Luis Reservoir (1963–1968) and the Oroville Dam (1957–1968) ensured that millions of individuals were provided with water.

Water distribution and production soon came under the joint control of the water authorities, who are often accused of being bureaucratic and secretive. Their advanced system provided state control and federal subsidies, enabling public funds to be used to finance infrastructure schemes that benefited coastal urban centers at the expense of dry, rural regions.

Most individuals think that this system is a form of socialism, which means the government regulates how items such as freshwater are produced, distributed, and sold. In California, the government plays an important role in the supply of water, which further substantiates this fact. This regulation has allowed the state to control the development of the region by allocating water to cities and agriculture while maintaining sustainable use. But others point out that state and federal domination of water rights has created issues like inefficiencies, corruption, and environmental destruction. In California, agencies like the Los Angeles Department of Water and Power (LADWP) and the California Department of Water Resources (CDWR) control vital water supplies, so they have considerable sway over the economic and political future of the state.

Water management is now in the hands of just a few appointed authorities, and this has resulted in huge disparities in access to water. In the past, communities used their own natural water sources. Now, they have to follow state and federal policies, even if these rules are harmful to them. They have little power to fight these decisions. This situation has led to discussions about fairness in how water is shared. Many rural communities feel ignored, while cities use a lot of water and often waste it.

The Example of Private Ownership Versus A Communist Farm

The example of a communist-run farm versus privately-owned farms is a long-accepted case study of the benefits of capitalism versus socialism or communism. In such examples, the primary difference is motivation to work. If you are a worker on a communal farm and work for 10 hours per day, your share of the farm's production is no greater than that of a worker who chooses to work for 2 hours per day. A worker/owner of a private farm enjoys the true fruits of their labor. Whether through ownership of a farm or simple ownership of a family home, ownership of real property is equivalent to buying into owning a piece of America. The ability of this generation and the next to join the American dream and participate in the great potential of this nation is dependent upon these generations' ability to purchase and specifically not rent the dwelling in which they live. This is an indisputable economic fact that currently has a negative impact on the US economy. Increasing the number or percentage of renters who rent "affordable housing" is a socialist-to-communist concept equivalent to slavery or at least indentured servitude.

The motivation for a homeowner to pay their mortgage and reap the rewards of such an **investment** is simply far greater than the motivation of a renter to pay their rent. A tangible benefit exists in ownership.

The creation of large infrastructure projects like the California Aqueduct and the Central Valley Project further entrenched the dominance of the state and federal governments in water distribution. While these projects have allowed for the expansion of cities along the coast, they have also resulted in environmental degradation, reduction of farm production, and the destruction of entire ecosystems. The environmental impacts of large-scale water projects have raised questions about the sustainability of the current system and whether it can meet the needs of future generations.

The Complexities of Water Redistribution in California's Infrastructure

(Major Aqueducts in the Southern Half of CA)

The image of California's water redistribution system serves as a powerful metaphor for understanding the state's massive water supply network and its challenges. The red lines indicate artificial water conveyance systems. All shades of blue lines indicate natural water conveyance systems. What happens to the volume of water in the natural system if you replace the artificial flows of such aqueducts with coastal desalination? What do you do with the constant flow of water provided by such aqueducts if you replace the demand for freshwater with coastal desalination? Yes, this can be easily done with existing technology. The political "hang-up" exists because executing such a strategy will solve the problem. Employees within the system have zero incentive to solve the problem. By examining the system, one can draw an analogy similar to a simple business solution of supply and demand. Water, like blood, flows through a vast network of "arteries" and "capillaries," carrying essential resources to where they are needed. However, the pressure placed on this system especially during emergencies like wildfires—reveals the strain that such a system is under.

During California's recent fires, we witnessed the water system's capacity being tested. As thousands of fire hoses were turned on to battle the flames, the existing infrastructure was pushed to its limits. While the larger "arteries" or aqueducts that carry water from the Central Valley to the high deserts and coastal cities functioned well under strain, the "capillaries"—the smaller pipes that bring water directly to homes became overloaded. The abrupt increase in water demand due to firefighting and normal use revealed an imperfect yet vital comparison: similar to a human body facing cardiovascular strain, the system continued to function but showcased the vulnerability of its limits in extreme situations. The water kept flowing, yet there were instances where the system wasn't meant to operate under that level of stress.

The Role of Hydroelectric Power in California's Water System

An additional layer of complexity in this infrastructure is the dependence on hydroelectric energy for water transport. While hydroelectric plants contribute significantly to power production, they fall short of meeting the energy demands needed to move large volumes of water to places like Los Angeles. The Pastoria Energy Facility, a natural gas turbine plant, is an example of how energy production is currently supplemented with non-renewable sources to meet these demands.

The question arises: why hasn't a broader shift to renewable energy, such as wind or solar, been proposed as a solution to power the water redistribution system? The simple answer is that wind and solar energy cannot **ever, ever, ever, ever** become a solution due to the laws of thermodynamics alone, such concepts are simply physically implausible. In order to provide the on-demand power required to move water in the quantities needed for cities like Los Angeles,

President Trump is correct. An "all the above" solution to power demands is **ABSOLUTELY REQUIRED.** In short, the "nuclear option" related to the equation E= mc^2 is the primary source of power required to provide the water needs of this largely overdeveloped region. Given the massive costs involved in transitioning to renewable energy at this scale, coupled with the limitations of current technology, it's clear that a significant energy shift would be a monumental undertaking.

The Aging Infrastructure

California's water system, much like the state's infrastructure at large, is aging. The system of aqueducts, canals, and pumping stations—built more than a century ago—was designed to handle different circumstances. As the state faces increased demand, especially from a growing population and exacerbated by climate change, the strain on these systems grows.

The situation is similar to a body's blood system when stressed. It's still working, but it's clear it needs urgent repairs and upgrades. Take the housing problem across the United States, for instance. Rebuilding affluent areas like Pacific Palisades creates homes for wealthy people, but there are greater concerns about providing affordable homes for everyone in the future. We need to consider these infrastructure issues, involving both housing and water, as part of a bigger plan for development. After the fires, we have a chance to rethink how we use resources and improve the strength and reliability of both the housing and water systems.

Currently, water systems mostly benefit cities near the coast, which often leaves rural and farming areas with less support. To improve this, we need to develop fair systems that make water and housing affordable and accessible everywhere in California.

The Complexities of the California Delta System

As we learn more about managing water in California, we can understand the complex Bay-Delta system better. Since 2011, experts have gathered a lot of data and used models to see how water moves in and out of this area. The Bay Delta is vital because it provides drinking water to the San Francisco metro area and supports much of California's agriculture. People worry about their health because the mixture of freshwater and saltwater is delicate. This balance gets upset when too much water is taken out, especially if the flow into the sea drops below 4,500 cubic feet per second (CFS), which can lead to problems.

Prior to the construction of any dams or water diversions, the Delta delivered approximately 30 million acre-feet of water to the ocean annually. Today, this has been reduced to roughly 20 million acre-feet per year. This reduction has made San Francisco Bay's water saltier and posed problems for water export. The individuals in charge of water pumping need to determine how to expand exports without harming the environment.

To address this, the "Big Gulp, Little Sip" strategy is suggested. This strategy entails exporting more water without compromising the environment. Rather than merely increasing pumping volume, it maximizes when and how it pumps. It takes into consideration tidal impacts and seeks to maintain balance in the Delta ecosystem. Through this strategy, we may export more water without allowing saltwater to invade the freshwater zones.

The federal and state water channels can carry 10,500 cubic feet of water per second at the moment. By improving how these channels operate and following better environmental practices, we could increase the amount of water sent by 2 to 3 million acre-feet each year. This additional water supply would be enough to support both farming and city needs. However, achieving this would require a significant investment to build new infrastructure, like reservoirs located south of the Delta. It's important that any solutions we implement do not harm the environment in any way.

Modern Challenges and the Future of Water Distribution

Lately, the debate over water rights and distribution has been a big issue, particularly in California, where severe droughts and climate change are making things worse. A major subject of discussion is the building of new pipelines, like the one planned from the Imperial Valley to San Diego. This debate highlights the battle over who should control water resources. As a nation, we can't keep taking more freshwater away from farms to support city growth because it leads to higher food prices. The planned pipeline aims to move more water from the Colorado River to supply the growing urban areas in Southern California, which only increases the conflict between city regions and rural farming areas.

Even with current challenges, there's hope that modern technology and new infrastructure can help fix some of the problems caused by past poor management of water resources. Desalination and water recycling improvements can significantly help. For many years, more efficient irrigation systems have been developed and used. We have now reached a stage where making more changes to irrigation costs a lot without much benefit. Instead, investing in desalination or water recycling gives better returns and is more cost-effective.

The Call for Strategic Infrastructure Investment

The water infrastructure in California is reaching a turning point. Many of the state's reservoirs, pumping stations, and aqueducts have been in operation for more than a century, but they are nearing the end of their useful lives. The weaknesses in this system have been exposed by the stresses of population increase, climate change, and rising water demand. A reason "Why the Hydrants went Dry" is that the California water system uses too much imported water to try and supply its coastal residents with fresh water. Proper cost controls and distribution of these costs to the appropriate entities appear to be beyond the capabilities of the California legislature and bureaucracies. Significant investment is required if we are to continue supplying water to the busy coastal cities as well as the rural heartlands.

We should understand that we can't make these investments by themselves. It's important to consider water, energy, housing, and environmental protection together. We need to take a comprehensive look and focus on sustainability, fairness, and long-term strength. By learning from the past 100 years and using a well-thought-out, fair plan, we have the best chance to tackle future challenges effectively. We must also recognize that California is but one of many states within the Arid Region that face these issues.

Chapter 3

Urban Sprawl and Its Impacts on Infrastructure

California is just one of 7 states that receive water allocations via the Colorado River Compact. California is the largest user of water in these states, and the people of this great state have experienced the largest shortcomings related to water bureaucracies failing the residents. With time, physical water flows in the Colorado River have declined from an average of 14.5 million acre-feet per year (1906–2020) to 12.5 million acre-feet per year since 2000. However, recent research suggests flows could rebound to between 13.5 and 14 million acre-feet on average between 2026 and 2050[16].

Colorado River Water Allocations		
User	**Amount (MAF)**	**Share**
United States	**15.0**	**90.9%**
CA	4.4	26.7%

[16] Mullane, S. (2024, May 15). There may be a wetter future ahead for the Colorado River, new CU Boulder research says. Water Education Colorado. https://watereducationcolorado.org/fresh-water-news/colorado-river-drought-cycles-cu/

CO	3.88	23.5%
AZ	2.8	17.0%
UT	1.72	10.4%
WY	1.05	6.4%
NM	0.84	5.1%
NV	0.3	1.8%
Mexico	**1.5**	**9.1%**
Total	**16.5**	**100%**

California state's reliance on complex infrastructure and limited natural resources has shaped its current water crisis. The state requires a substantial water supply, as it cultivates nearly 99%[17] of the nation's fruit and nut production. The current case of wildfires has further intensified the need for a dependable water supply, creating a delicate balance between managing finite resources and sustaining a thriving population.

The environmental impacts of over-extraction and mismanagement, as detailed in the previous chapter, are now becoming more evident, with long-term consequences that threaten the state's ecosystems, the safety of its residents, and future water availability.

In response to these mounting pressures, California is turning to innovative solutions, seeking to address the fundamental need for water across communities and industries. This chapter explores some of these groundbreaking efforts, from the use of Earthly Water Reclamation Systems to advanced technologies that promise more efficient water management. California's escalating housing crisis, paired with the rapid expansion of data centers, further complicates the availability and accessibility of water. Meanwhile, the innovative potential of seawater desalination and NASA's cutting-edge water recovery systems aboard the International Space Station provide a glimpse into futuristic solutions. Through these technologies and initiatives, California plans to build a path toward sustainable water use and recovery.

---

[17] Rance, H. (n.d.). 19 Crops That Are Grown Only in California. CA GROWN Blog. https://californiagrown.org/blog/only-in-california/

Unfortunately, California has completely failed to explore common, affordable, and practical solutions, instead favoring the most expensive and ill-conceived projects imaginable. The same officials responsible for these decisions also promoted the failed multi-billion-dollar high-speed rail project between Los Angeles and San Francisco. Similarly, initiatives like the Bay Delta Conservation Plan (BDCP) and other large-scale projects have yielded comparable disappointing results.

Water Reclamation Systems

Water reclamation technology has evolved over the last 20 years to meet the demands of a growing population. These systems can use aquifers and riparian streams, which are capable of purifying wastewater to meet drinking water standards. As the next frontier of water management, advanced water reclamation technologies may help us achieve higher levels of water recovery and purification, much like the systems NASA has developed on the International Space Station (ISS). NASA has made significant strides, recycling as much as 98%[18] of the water used aboard the ISS, which is a crucial model for closed-loop water systems in space and could provide insights for creating more efficient water recovery processes on Earth.

Reclaimed Water & Recharge

Various towns have successfully implemented water reclamation strategies. One such achievement is the recovery of 65% of the water served, which is treated and stored below ground for future use. This recovered water can be pumped from remote locations within the same basin, ensuring a sustainable and efficient water supply, even for distant areas.

Arizona's Department of Water Resources (ADWR) has developed a "Credit" accounting system to track the amount of water that can be pumped for recharge, supporting long-term water conservation efforts. This method helps manage and preserve the region's water supplies.

[18] ISS: Astronauts have recycled 98% of water from their pee, sweat and breath. (2023, June 27). BBC. https://www.bbc.co.uk/newsround/66017072

Advanced Wastewater Reclamation Plants

Several innovative wastewater reclamation plants have been established to further support sustainable water management. Examples include:

- Hyperion Advanced Water Purification Facility in Los Angeles, which is working towards recycling 100% of its wastewater that makes it to the plant, but some do not.
- Donald C. Tillman Water Reclamation Plant, also in Los Angeles, which is central to the city's efforts to reduce dependence on procured water and ensure a drought-tolerant water source. These plants not only play a vital role in delivering water for irrigation and industrial use but are also crucial for indirect potable reuse (providing treated water for drinking after advanced purification).

Case Study: Prescott Valley, Arizona

Prescott Valley, Arizona, has employed advanced water reclamation techniques that recover a significant portion of wastewater. The town recovers 65% of the water provided to homes, businesses, and other consumptive uses. More specifically, more than 65% of the water served is channeled into sewage lines, where it undergoes advanced water reclamation processes. These processes reclaim about 90% of the water that enters the system, with only about 10% of the water lost in the form of the moisture content of the "cake" left behind after filtration. This "cake" is removed from the aquifer and sequestered, ensuring it doesn't percolate back into the groundwater supply.

In Prescott Valley, it's estimated that 40-50% of the water that flows through tap water systems comes from reclaimed sources, a clear indication of how effective the water reclamation system is. Such initiatives show that the water reclamation technology can already support a significant portion of our water needs.

Similarly, our nation's growing housing demand and the need for large-scale data centers also pose significant challenges for water management. As new developments rise to meet these needs, integrating water reclamation systems into modular housing systems such as Fullstack Modular housing or Z modular housing can help reduce the pressure on local water supplies.

Modular houses are prefabricated, scalable housing units designed to address the affordable housing crisis across the nation and beyond. These modular units are built off-site and then assembled on-location, offering quicker construction and reducing overall costs.

For residential and commercial infrastructures, these modular housing systems are coupled with cutting-edge water reclamation technologies. They might therefore be used as models for locations that experience persistent water scarcity, especially in arid regions.

Water Recovery in Housing Development

Water recovery system integration and careful planning are necessary when constructing sustainable housing in areas with limited water resources. Advanced modular homes placed close to aquifers can help optimize water recovery rates while maintaining livability in communities. For example, housing projects that have water reclamation systems installed on-site or nearby, such as those offered by GWT, may see recovery rates of 80–90%[19]. With such advancements, these systems can lessen dependency on outside water sources and ease the stress on overworked water infrastructure.

Integrating Geothermal Energy in Data Centers and Housing

The integration of sustainable technologies such as geothermal energy for heating and cooling systems can contribute to making new developments more energy-efficient. Data centers, a critical component of modern technological infrastructure, consume vast amounts of power—about 40%[20] of which is used for cooling. By utilizing geothermal temperature moderation or waste heat from natural gas turbines, these systems could provide cost-effective and efficient cooling and heating. This will further reduce the environmental footprint of water and energy use in new developments.

---

[19] Pros and Cons of High Recovery Reverse Osmosis Systems. (n.d.). Puretec Industrial Water. https://puretecwater.com/resources/pros-and-cons-of-high-recovery-reverse-osmosis-systems/

[20] Zhang, X., Lindberg, T., Xiong, N. (et al.) (2017). Cooling Energy Consumption Investigation of Data Center IT Room with Vertical Placed Server. Energy Procedia. 105. 2047-2052. 10.1016/j.egypro.2017.03.581

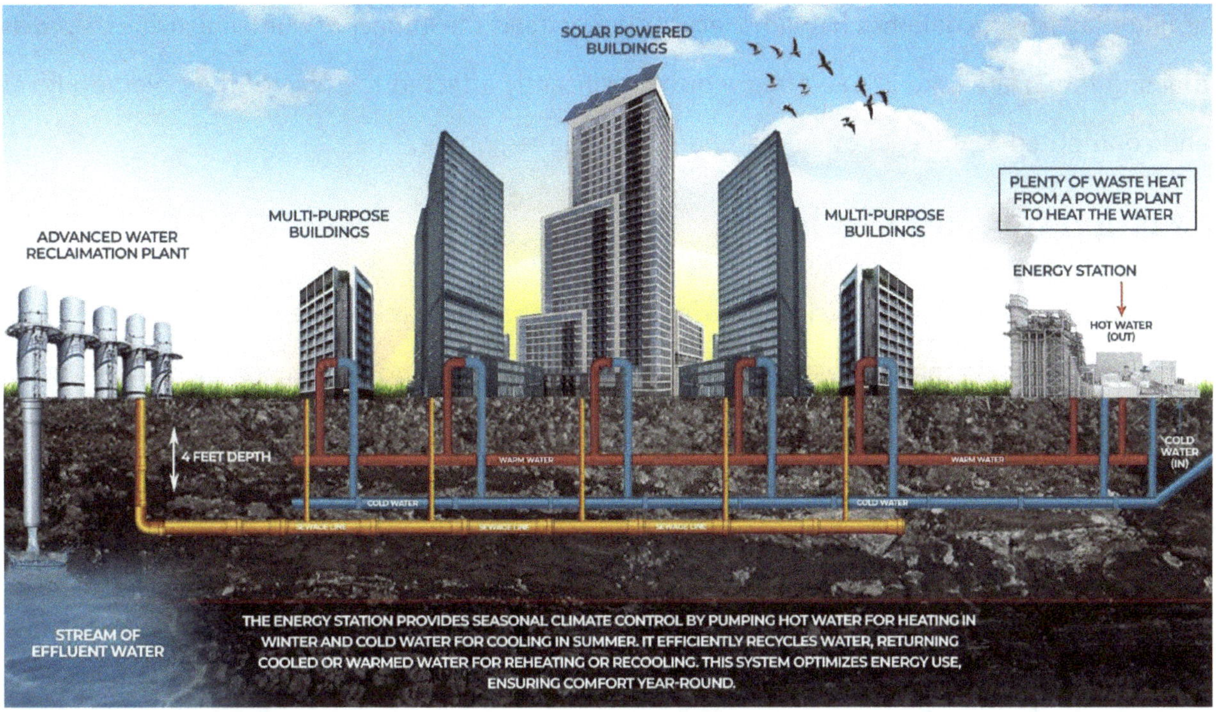

(Model of Sustainable Housing and Data Centers)

Colocation of data centers, and new Freedom Cities would allow Data Center developers and affordable housing developers to share grading and utility extension costs. Depending on the extent to which the federal government wishes to subsidize the creation of affordable housing on federal lands, the land itself should be inexpensive or free. A city built in this manner with a footprint of 1 square mile could house 30-40k residents. Having AI as a roommate living in the basement will provide a literal and figurative foundation for the city and its economy.

Satellite Technology for Water Management: The GRACE Mission

One notable technology in water management involves the use of satellite systems like the GRACE (Gravity Recovery and Climate Experiment) mission. This system employs satellite pairs that track the variations in Earth's gravitational field to measure the amount of water stored in aquifers. By using this technology, we can more accurately monitor and manage underground water resources, enabling better decision-making regarding water allocation. This technology can be particularly helpful in areas where groundwater is the primary source of water. It's also worth noting that satellite data has helped scientists understand the impact of human activities, such as draining the Tulare Lake in California, which significantly affected the flow of the Colorado River and contributed to the environmental changes in the region.

Artificial Water Cycle

The Moon and Mars, unlike Earth, are devoid of natural rain cycles. The prospect of creating a human presence on these celestial bodies necessitates the creation of an artificial "water cycle." Similar to the desert world of *Arrakis* in *Dune* (but far colder), these planets are hostile to water as we know it. But how will humanity meet its water needs in such an environment?

One possibility that might be more efficient than transporting water from Earth is harnessing ice from the outer rings of Saturn or the asteroid belt. Elon Musk, among others, is looking into the possibility of sending drones to Saturn's ice-rich outer rings to transport large icebergs to fuel space settlements. This process would conserve angular momentum and use materials found elsewhere in the solar system rather than rely on Earth's limited water supply.

While that may seem outlandish, it's essential to consider how water systems will need to evolve in space. Water reclamation and recycling will be a crucial part of future off-world colonies. In space, the recovery of water will need to be close to 100% efficiency, given the limited resources. The technology currently in place on Earth—such as advanced water reclamation systems—could become the foundation for these extraterrestrial systems. For instance, NASA has accomplished remarkable water recycling rates on the ISS, recovering 98%[21] of water from perspiration, urine, and other sources.

Groundwater Depletion of the Central Valley

To observe how water reclamation and sustainability impact our world, it's necessary to look at areas where water management is failing. The Central Valley in California is a prime example. It used to have a large body of water, Tulare Lake. In the late 1800s, a landowner, J.G. Boswell, drained the lake to turn the lakebed into agricultural land. This transformation was good for

---

[21] ISS: Astronauts have recycled 98% of water from their pee, sweat and breath. (2023, June 27). BBC. https://www.bbc.co.uk/newsround/66017072

farming, but created a number of issues. It significantly decreased the groundwater, which resulted in the substantial reduction of the region's water resources. This scenario reveals the impact of utilizing water unsustainably.

The consequences of the misuse of water can be observed in the low water levels of the Colorado River and the dwindling Great Salt Lake. These are problems that have a long-term basis, reminding us that we should use water conservatively. Regrettably, numerous regions of California and the western US continue to experience difficulty with this. There is hope, however. New technology, such as desalination plants and advanced water recovery systems, may provide the means to solve these challenges. For example, big desalination plants, such as those spanning 449 acres in Dubai, could potentially supply sufficient water to areas that experience droughts on a regular basis. These plants produce up to 659,726 acre-feet of water annually, presenting a possible solution to the water supply issues of the region.

Water Efficiency and the Challenge of Aquifer Depletion

As urban populations grow, the challenge of managing water resources becomes significantly more pressing. The volume of freshwater resources accessible for irrigation has notably diminished due to aquifer depletion, particularly in regions such as California's Central Valley. Previously a reliable source of water, groundwater is now being consumed at unsustainable levels.

To tackle this issue, certain areas are adopting advanced water technologies, like those implemented in Prescott Valley. By investing in systems for water reclamation that treat wastewater, communities can restore significant portions of their water resources. Nonetheless, the primary concern is the quantity of this reclaimed water that can be safely reintroduced into the environment or utilized in agriculture.

In places like Idaho, advanced wastewater reclamation plants could support local communities by offering a more sustainable model for water use. These plants can recycle water for irrigation, landscaping, and even agricultural uses, reducing the strain on local water systems. However, there will always be questions about how much-reclaimed water residents are willing to return to rivers and how much they prefer to keep within their local systems.

California Water Efficiency Programs

California has developed several programs and initiatives to help urban and agricultural communities optimize water use, ensure reliable supplies, and address the long-term impact of climate change.

Failure to fill local reservoirs in order to eliminate evaporation from the water surface is an example of government efforts to achieve greater efficiencies and place revenue above public safety. Central to these efforts are the state's various funding and technical assistance programs, which target water conservation, efficiency improvements, and long-term sustainability. Most of these programs are prime for government waste, fraud, and abuse.

One such initiative is the Water Use Efficiency Financial Assistance Program (WUEFAP). This program provides funding to both urban and agricultural agencies, helping them implement projects designed to improve water efficiency, achieve substantial water savings, and reduce operational costs. Beyond conserving water, these projects also contribute to enhancing water quality, cutting energy consumption, and providing critical environmental benefits. By targeting a range of sectors, WUEFAP helps California communities better cope with both seasonal and chronic water shortages while advancing long-term resilience. Projects like this generally have an extremely high cost-to-benefit ratio. The WUEFAP is a small part of Big Brothers' propaganda campaign initiated to reduce water availability to the world-class farmland in the Southern Central Valley. Like the high-speed rail project from LA to San Francisco, state-run programs like this **always** lead to cost overruns, construction delays, and end products that **never** produce the promised results. From a national security perspective, food production security is a real concern. Reductions in food production in the Southern Central Valley will lead to the burning of more rain forests in Central and South America. People will always need to eat. Growing sufficient food on US soil to keep the grocery store shelves full is a matter of national security, similar to producing sufficient energy, steel, data centers, chips, and water to quench demand from a growing population. The answer here is that the cost-benefit ratio relative to the US economy requires the delivery of more water. This can be done in ways that are environmentally and fiscally sound. Programs to increase water efficiency in the agricultural sector have already run their course as they relate to the concept of diminishing returns. The continuance of such projects is like throwing money at a problem that this strategy will never solve. Repeating the same experiment and expecting a different result is the definition of insanity.

Similarly, the Small Suppliers Water Conservation Assistance Program serves smaller water suppliers and communities that face unique challenges in managing water resources effectively. Specifically aimed at suppliers serving between 15 and 2,999 connections, this program addresses the difficulties these entities often encounter in acquiring funding and implementing drought and conservation programs. The assistance offered is broad, covering technical, managerial, and financial support, as well as demand management implementation.

Notably, the program identifies particularly vulnerable suppliers based on water shortage risk assessments, ensuring that those most in need receive assistance. There isn't an official application process, but eligible small water suppliers get contacted directly by program representatives to offer assistance. In California, it is key to monitor and analyze water use and conservation. For this purpose, the state uses the WUE Data Portal. This is an online platform where urban and agricultural water suppliers can send in and update important documents. These documents are on urban and agricultural water management. They are water loss reports and farm-gate delivery reports. The portal is open and accessible, and researchers, policymakers, and the public can view and review the information provided without a login ID.

The Urban Water Management Plan (UWMP), a comprehensive document developed in accordance with state law every five years, is a vital component of California's water management plan. Beyond defining a water agency's activities, water supply, and demands, the UWMP also outlines its strategies for resilience and drought risk management. Beyond being legally mandated, this plan is an important tool for long-term water management planning.

In developing and adopting the WSCP, agencies are required to go through a well-established public process so they are being transparent and consulting with the people they serve. With the inclusion of a set of contingency measures, the WSCP assists communities in staying flexible and ready for the uncertainty of California's water supply, which is ever-increasingly affected by climate change impacts.

Addressing the Affordable Housing Crisis and Data Centers

Urban development and water are issues that affect more than simply single-family houses. With rapid population growth, the demand for affordable housing always grows. Coupling new data centers and power sources with new freedom cities would have a mutual cost benefit if redundant sources of power are available to power the data center. The need for new Data Centers and the power required to run them can serve as an example of one example of the industry required to create the good-paying jobs needed to make housing affordable. These data centers consume vast amounts of energy, about 40% of which is spent on cooling, especially when data centers are built in hot environments. There are no known engineering principles that would prevent using the same footprint to build affordable housing on top of Data Centers. This includes security issues. Such an approach would actually reduce the cost of construction for both.

One potential solution lies in combining energy efficiency with water conservation. Geothermal systems could be used to provide both heating and cooling for new urban areas, significantly reducing the amount of energy consumed by data centers. Geothermal energy could even provide heating and cooling for modular housing systems, improving sustainability and reducing the need for external power sources.

The use of natural gas or small modular reactors (SMRs) could support clean energy generation, and the waste heat from these systems could be used for water and air heating in the cooler environment within the Upper Colorado River Basin and the Snake River Basin in Idaho, all on federal lands. This type of integrated infrastructure would ensure that power and water usage are both efficient and sustainable. Moreover, newly designed homes in the Freedom Cities feature AI systems integrated into the basement, effectively reducing utility costs for both humans and machines.

The Future of Water Supply

To manage water on a global scale, satellites and remote sensing technologies will become even more important. The Grace Satellite System, for example, uses two tethered satellites to measure the gravitational pull of water stored in aquifers. Researchers can use these data to monitor changes in water storage and improve the management of water resources both locally and globally.

We can better depict where water is being used, how much is available, and where it is being depleted with the usage of this technology. Such insights will be invaluable as we move toward more sustainable water practices, both on Earth and in space. Even with these advanced technologies, nothing can replace having a small number of "boots on the ground."

Water Cycle and Runoff in the Southwest

Recent studies show that irrigation practices in California's Central Valley impact the larger southwestern US water cycle. Through increased evapotranspiration (ET), irrigation contributes significantly to atmospheric moisture, which in turn strengthens the regional hydrological cycle. The research, led by Min-Hui Lo and James S. Famiglietti[22], demonstrated that irrigation in this region leads to:

- A 15%[23] increase in summer precipitation, benefiting agricultural regions.
- An increase in Colorado River streamflow by 30%[24], with a resulting boost in water available for California and surrounding states.

This effect was particularly evident when moisture from the Central Valley was transported over the Sierra Nevada and contributed to monsoon systems, increasing rainfall in other states. However, this also led to more intense seasonal weather patterns, including monsoons, which exacerbated the severity of storms and flooding.

---

[22] Lo, M., & Famiglietti, J. S. (2013). Irrigation in California's Central Valley strengthens the southwestern U.S. water cycle. Geophysical Research Letters, 40(2), 301–306. https://doi.org/10.1002/grl.50108
[23] Ibid.
[24] Ibid.

The King of California: J.G. Boswell's Water Empire

J.G. Boswell built the largest agricultural empire in the US, using water in innovative ways to transform large swathes of land into highly productive cotton fields. Known as the King of California, Boswell's empire involved draining one of America's largest lakes and utilizing advanced water management techniques. His ability to control water supplies to sustain extensive agricultural operations served as the foundation for his empire.

Seawater Desalination: The Path to a Sustainable Water Supply System

California is the only state in the union where large-scale desalination is practical and required for future growth in a safe and efficient manner. Some progress has been made. However, the tiny country of the United Arab Emirates (UAE) desalinates over 2.25 million acre-feet of water per year. No significant impacts on the marine environment off the coast of the UAE have ever been documented. The UAE, Saudi Arabia and others have been desalinating ocean water on an ever-growing scale for decades. The example below of the Claude "Bud" Lewis desalination plant in San Diego is a good start. This plant only desalinates about 50,000 acre-feet of water per year[25]. This good start needs to increase in scale and number if California is to have a sustainable future.

The Claude "Bud" Lewis Carlsbad Desalination Plant in San Diego County, California, is one of the most well-known instances of extensive desalination projects in the United States. This cutting-edge facility, which is the biggest and most technologically sophisticated saltwater desalination plant in the country, is essential to supplying the water needs of the area. On December 14, 2015, Poseidon Water and the San Diego County Water Authority officially dedicated the Claude "Bud" Lewis Carlsbad Desalination Facility. Over 500 officials and advocates from across California participated in the event, marking the end of a $1 billion effort focused on diversifying and securing the region's water supply.

[25] Hartig, T., & Jeyaram, S. (n.d.). Reverse Osmosis in Desalination Plants. University of Pittsburgh. https://sites.pitt.edu/~budny/papers/8117.pdf

The facility, which is included in a broader infrastructure initiative featuring a 10-mile pipeline and several enhancements, satisfies around 10%[26] of the region's water needs. Or, it can deliver about one-third of the water produced in San Diego County. This supply is distributed to over 22 local water agencies, which serve 3.3 million people and support a $253 billion economy[27].

In order to have a real impact on water availability in California, we will need a far larger capacity to desalinate and integrate the water distribution system. A substantial number of the existing aqueducts actually cross each other, and there are no existing means of transferring this critical resource. The days of each public bureaucracy owning its own supply system without the ability to distribute temporary overabundance into other conveyance systems must come to an end. Many of the existing aqueducts cross each other. This provides a clear and inexpensive opportunity to integrate the system.

Integrating multiple desalination plants capable of desalinating a volume of water similar in scale to that of the small country of the United Arab Emirates at 1.5 million acre-feet per year would be a great step towards sustainability. Please note that the entire state of California only desalinates about 150 thousand acre-feet per year (10 times less than the UAE).

This integration of the old system and desalination provides a long-sought abundance of water and a redundant water supply for emergencies.

The Water Purchase Agreement

The Water Purchase Agreement was a crucial component of the project's financing and operational framework. The agreement transferred the risks associated with the design, construction, and operation of the desalination plant to Poseidon and its investors. It also

---

[26] Carlsbad Desalination Plant Celebrates 5th Anniversary. (2020, December 14). Water News Network. https://www.waternewsnetwork.com/carlsbad-desalination-plant-celebrates-5th-anniversary/

[27] Seawater Desalination The Claude "Bud" Lewis Desalination Plant and Related Facilities. (2025, February). San Diego County Water Authority. https://www.sdcwa.org/wp-content/uploads/2020/11/desal-carlsbad-fs.pdf

transferred the risks associated with the construction of the 10-mile pipeline[28] that delivers desalinated water from the plant to the Water Authority's Second Aqueduct in San Marcos.

Under the terms of the agreement, the Water Authority has the option to purchase the plant starting in 2026 and the right to acquire it for $1 in 2046[29], thus enabling public ownership of the plant, intake, and discharge facilities, and the long-term lease with NRG (which owns the site).

Financing Terms

To finance the desalination plant and pipeline, the Water Authority and Poseidon secured tax-exempt construction bonds in December 2012. The financing closed at a favorable interest rate, bringing the project's financing costs $200 million below the Water Authority's initial projections. This efficient financing structure contributed to the project being recognized as the "North American Water Deal of the Year" for 2012 by Project Finance, an international trade publication. The publication noted that the $734 million[30] bond issue "could serve as a useful template for public-private partnerships in the water industry, particularly for seawater desalination projects."

Water Purchase Price

As outlined in the Water Purchase Agreement, the price of desalinated water for the 2019 fiscal year ranged from $2,302 to $2,559 per acre-foot[31], depending on the volume purchased. The first 48,000 acre-feet of water purchased each year covers both fixed and variable costs of the project. The Water Authority also has the option to purchase up to an additional 8,000 acre-feet per year at a lower rate based only on the variable costs of water production. The typical monthly cost per household is about $5, which is at the low end of the Water Authority's forecast in 2012.

---

[28] Water Purchase Agreement for Seawater Desalination Nearing Completion. (2012, August 9). San Diego County Water Authority. https://www.sdcwa.org/water-purchase-agreement-for-seawater-desalination-nearing-completion/

[29] Ibid.

[30] With Financing Complete, Poseidon Inks Deals for Construction, Operation of Carlsbad Desalination Project. (2012, December 24). Poseidon Water. http://www.poseidonwater.com/news-and-events/with-financing-complete-poseidon-inks-deals-for-construction-operation-of-carlsbad-desalination-project

[31] Desalination. (n.d.). Water Education Foundation. https://www.watereducation.org/aquapedia/desalination

Annual price increases for water are capped at 2.5% for inflation, which compares favorably with the 7%[32] annual increases in treated water rates imposed by the Metropolitan Water District of Southern California between 2008 and 2018. Additionally, Poseidon is permitted to increase prices to accommodate industry-wide changes in laws or regulations, with cumulative increases capped at 30% over the 30-year term of the agreement[33].

The Plant and Desalination Process

The Carlsbad Desalination Plant is located on six acres of public utility-zoned land next to the NRG Encina Power Station on the Agua Hedionda Lagoon in Carlsbad. The facility was constructed at an estimated cost of $537 million[34]. The desalination process itself is based on a reverse osmosis system designed by IDE Technologies. Seawater is pumped to the plant, where it undergoes a sand/anthracite filtration process to remove suspended particles before being pushed through reverse osmosis membranes that remove salts and other dissolved particles. Essential minerals are added back to the water before it is distributed to the regional aqueduct system.

As of 2019, the Encina Power Station was scheduled to shut down and be replaced by a more modern power plant. This change will lead to the reconfiguration of the desalination plant's seawater intake system to comply with new environmental regulations. Poseidon will also take responsibility for dredging the lagoon to maintain the facility's environmental integrity.

Now we know that it costs just over $10,000 to build 1 acre-foot per year of desalination capacity, and running the plant costs about $2,500[35]. This sounds like an expensive means of providing water to a large city. For a real comparison, we would need to adjust the cost of building the

---

[32] City News Service. (2022, May 26). San Diego County Water Authority Proposes 5.2% Increase, Citing Inflation and Energy Costs. Nbcsandiego. https://www.nbcsandiego.com/news/local/san-diego-county-water-authority-proposes-5-2-increase-citing-inflation-and-energy-costs/2957211/

[33] Carlsbad Seawater Desalination Project Water Purchase Agreement. (2012, October 30). City of San Diego. https://www.sandiego.gov/sites/default/files/legacy/citycouncil/committees/wpitf/pdf/121030item3.pdf

[34] Seawater Desalination: The Claude "Bud" Lewis Desalination Plant and Related Facilities. (2020, February). San Diego County Water Authority. https://www.sdcwa.org/sites/default/files/desal-carlsbad-fs-single.pdf

[35] Seawater Desalination The Claude "Bud" Lewis Desalination Plant and Related Facilities. (2025, February). San Diego County Water Authority. https://www.sdcwa.org/wp-content/uploads/2020/11/desal-carlsbad-fs.pdf

California aqueduct (over $35 billion in today's money) and consider the cost of moving water uphill about 3,200 hundred feet in elevation. In short, the cost of building and operating a desalination plant is still a bit more expensive than building new conveyance systems, but it has now become "cost competitive."

Reverse Osmosis and Pipeline Integration

The desalinated water produced by the Carlsbad plant is transported through a 10-mile pipeline that connects the plant to the Water Authority's Second Aqueduct. This vital piece of infrastructure is owned by the Water Authority and was constructed at an estimated cost of $159 million[36]. Beginning in San Marcos and continuing to Vista and Carlsbad, the pipeline installation got underway in the spring of 2013. The majority of the pipeline was constructed in Carlsbad.

The Twin Oaks Valley Water Treatment Plant, located north of San Marcos, receives the desalinated water after it passes through the Second Aqueduct. Before being delivered to the area, the desalinated water is combined with the area's current drinking water supplies at the plant. The Water Authority's system integration of desalinated water has given the community a reliable and sustainable water source.

Reverse osmosis vessels are a crucial component of the desalination process at the Claude "Bud" Lewis Carlsbad Desalination Plant. After production, a new 10-mile pipeline carries the desalinated seawater to the Water Authority's Second Aqueduct system. The new pipeline and related control facilities, which include a major improvement to Pipeline 3, enable the safe and efficient transportation of desalinated water from Carlsbad to the Twin Oaks Valley Water Treatment Plant in San Marcos. To ensure that water quality standards are met, Twin Oaks mixes the desalinated water with its pre-existing treated water sources before distributing it. The water then flows south via Pipeline 4, which has also been strengthened to withstand the increased pressure brought on by the desalination process.

---

[36] Smith, E. J. (2022, June 17). Poseidon desal plant in Carlsbad needs $159M upgrade. Ratepayers will foot the bill. The San Diego Union-Tribune. https://www.sandiegouniontribune.com/2022/06/17/poseidon-desal-plant-in-carlsbad-needs-159m-upgrade-ratepayers-will-foot-the-bill/

For its technological prowess and environmental sustainability, the Claude "Bud" Lewis Carlsbad Desalination Plant received multiple distinguished awards:

- Global Water Intelligence, the publisher of leading industry periodicals, named it the Desalination Plant of the Year in 2016. The facility is considered one of the most sustainable desalination facilities in the world and received the award for its outstanding ecological and technical contributions to the desalination sector. But the plant is just too tiny to have a significant impact on San Diego's water budget, let alone the California State budget or the Arid West's.
- The Association of California Water Agencies (ACWA) presented the San Diego County Water Authority with the Clair A. Hill Water Agency Award for Excellence in 2017. This award was presented in recognition of the Water Authority's successful integration of Carlsbad Desalination Project resources into its comprehensive water management strategy, which prioritizes innovation and excellence in water resource management.

Water Authority Improvements

The Water Authority worked on improving the system so that they could add desalinated water to the regional supply more easily. They improved Pipeline 3, which plays an important role in the Second Aqueduct. Additionally, they made changes to the Twin Oaks Valley Water Treatment Plant to support this integration.

A five-mile section of Pipeline 3 was relined to handle the increased pressures resulting from the pumping of desalinated water uphill to the treatment plant. Additionally, the Twin Oaks plant underwent modifications to allow the desalinated water to be stored alongside imported drinking water supplies. These upgrades cost approximately $80 million[37] and ensured the seamless integration of desalinated water into the region's overall water distribution system.

[37] Seawater Desalination The Claude "Bud" Lewis Desalination Plant and Related Facilities. (2025, February). San Diego County Water Authority. https://www.sdcwa.org/wp-content/uploads/2020/11/desal-carlsbad-fs.pdf

Environmental Benefits and Technological Innovation

The most significant environmental benefit of a large-scale desalination plant (the Jebel Ali plant in the UAE produces about 650,000 acre-feet per year or 13 times the volume), like the Carlsbad plant on the California coast, will be the restoration of ecosystems affected by coastal cities' water extraction. The Claude "Bud" Lewis Carlsbad Desalination Plant was designed to meet stringent environmental standards set by state and local agencies, including the California Coastal Commission. The facility utilizes cutting-edge technology to recapture energy from the desalination process, significantly reducing carbon emissions. Additionally, it has contributed to the creation of extensive wetlands along the San Diego coastline, which support local fish populations and help maintain the ecological balance of the region.

According to Poseidon's Climate Action Plan, the plant should achieve net carbon neutrality within 30 years. By offsetting greenhouse gas emissions from the plant's operations, this ambitious goal is accomplished, making it the first significant infrastructure project in California to completely reduce its carbon footprint. The Carlsbad Desalination Plant significantly reduced its environmental impact by installing energy recovery technology and buying carbon offsets. The energy recovery systems save an estimated 116 million kilowatt-hours of energy each year, amounting to a reduction of around 42,000 metric tons of CO2 emissions[38]. This reduction is equivalent to the amount of greenhouse gas emissions of 9,000 passenger cars per year.

Poseidon has also attempted to restore regional ecosystems. In the hope to create a mosaic of coastal habitats that benefit fish and bird species, the business is actively rebuilding 66 acres of wetlands in San Diego Bay by digging and grading a former salt production pond. The establishment of public recreational paths is another aspect of this restoration.

Furthermore, Poseidon has assumed responsibility for the continued stewardship of the 400-acre Agua Hedionda Lagoon, committing to the restoration of 37 acres of wetland habitat in the area. Despite the clear advantages, the project faced significant legal challenges. It underwent a rigorous environmental permitting process and was subject to 14 legal challenges before

[38] Ibid.

construction could begin. However, the project successfully met all environmental compliance standards, and the final outcome has helped establish desalination as a viable, sustainable solution to water scarcity in the region.

Economic Benefits

The Carlsbad Desalination Project has provided substantial economic benefits to the region. During construction, the project supported an estimated 2,500 jobs and injected approximately $350 million[39] into the local economy. Now that the plant is operational, it directly supports 36 full-time employees and 124 indirect jobs. Poseidon also anticipates an annual direct spending impact of around $45 million related to the plant's ongoing operations across the region.

The Infrastructure: Pipeline and System Upgrades

The Carlsbad Desalination Plant is linked to the Water Authority's existing water distribution system through a 10-mile pipeline, which was constructed at an estimated cost of $159 million. Pipeline installation began in 2013 and included work in San Marcos, Vista, and Carlsbad, the latter being home to the majority of the pipeline's length. A new desalination plant on the shoreline of the nearby Camp Pendleton that could produce several hundred thousand acre-feet per year would require a more extensive integration into the existing system.

Once desalinated water is produced at the plant, it is delivered via the pipeline to the Water Authority's Second Aqueduct. The water is sent to the Twin Oaks Valley Water Treatment Plant, where it combines with the usual drinking water for the area. This ensures a stable supply for local residents. To handle the higher pressures from moving desalinated water uphill, the Water Authority improved its system by reinforcing a five-mile section of Pipeline 3. They also updated the Twin Oaks plant to store both desalinated and imported water. These upgrades, costing about $80 million, are crucial for smoothly integrating desalinated water into the local water system.

---

[39] Carlsbad Desalination Plant. (n.d.). Mott MacDonald. https://www.mottmac.com/en-us/projects/carlsbad-desalination-plant/

The Role of Desalination in a Diversified Water Supply

For over two decades, the San Diego County Water Authority has recognized seawater desalination as an essential component of a diversified water supply portfolio. The authority's strategy includes a combination of potable reuse, recycled water, groundwater, and independent water transfers. This diversification helps buffer the region against the risks associated with relying on a single water source, particularly during times of drought or other environmental challenges.

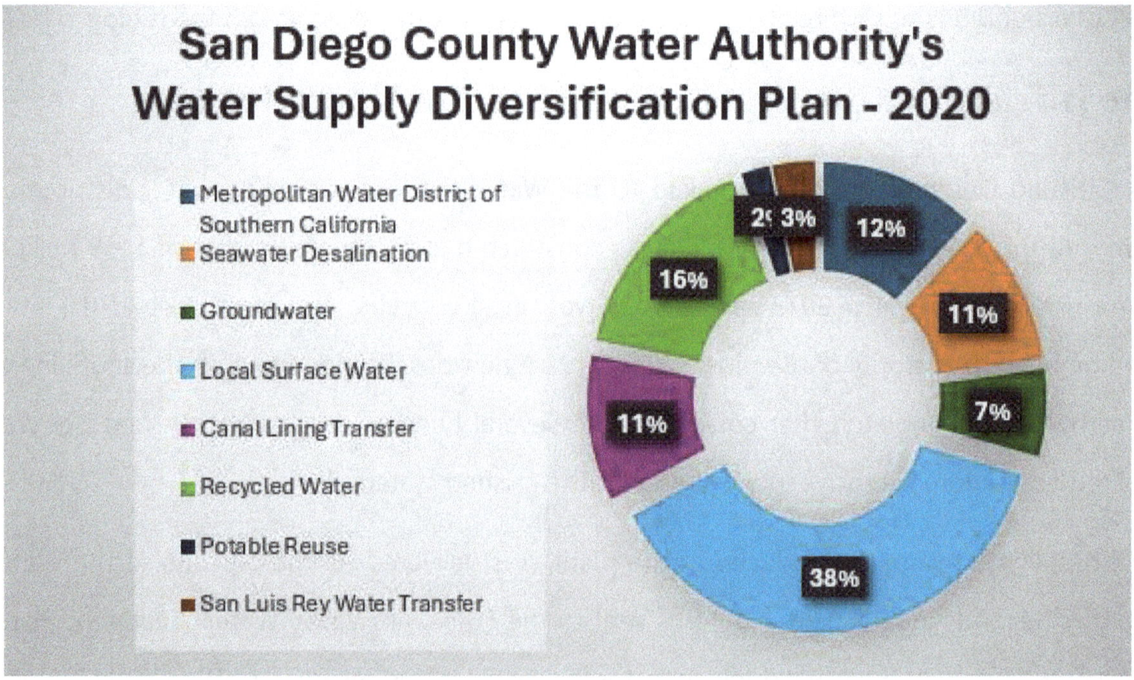

It should be noted that the "Canal lining Transfer" comes from adding a concrete lining to the All-American Canal. The lining was installed to prevent water from percolating into the local transnational aquifer shared with Mexico. Shortly after completion, the depth of water in the wells and south of Mexicali substantially increased. Much of this water would have percolated through the aquifer and surface downstream in the Colorado River Delta. A similar project in the San Francisco Delta would be unlikely to be approved.

Environmental and water advocates have long argued for the concept of regional self-sufficiency. This argument focuses on reducing or eliminating the 12% of water provided by the Metropolitan Water District.

As the Carlsbad Desalination Plant continues to provide reliable, locally controlled water, it offers a model for other regions grappling with water scarcity. The plant's success in providing 54 million gallons per day[40] has proven that desalination can be both technologically feasible and environmentally sustainable. Yet, one must acknowledge that desalination is not a complete solution to water shortages. It works best in conjunction with other water management measures, like water reclamation and efficiency measures.

NASA's Water Recovery Achievements on the ISS

NASA accomplished something remarkable on the International Space Station (ISS) by reclaiming 98%[41] of water, which means almost all of the water can be reused. This was accomplished with a water recycling system called the Environmental Control and Life Support Systems (ECLSS). The ECLSS is made up of several parts, including dehumidifiers, a Water Processor Assembly (WPA), a Urine Processor Assembly (UPA), and a Brine Processor Assembly (BPA). These technologies enable the transformation of astronaut sweat and urine into potable water.

Achieving a 98% recovery rate is essential for upcoming space missions, especially those venturing far from Earth, such as missions to the Moon or Mars. In these distant missions, getting new supplies from Earth is quite difficult. By recycling water efficiently, NASA can provide astronauts with a sustainable water supply without needing constant shipments from Earth.

This innovation in water recycling is important as it could enable long-term space exploration. It reduces the reliance on resources that would otherwise need to be sent from Earth, making long journeys into space much more practical.

---

[40] Claude "Bud" Lewis Carlsbad Desalination Plant. (n.d.). Kiewit Corporation. https://www.kiewit.com/projects/claude-bud-lewis-carlsbad-desalination-plant/

[41] ISS: Astronauts have recycled 98% of water from their pee, sweat and breath. (2023, June 27). BBC. https://www.bbc.co.uk/newsround/66017072

GRACE Satellite Data and Groundwater Management

NASA's GRACE satellite has provided important insights into how climate and irrigation practices impact water storage in the United States. Here's what was found:

- In wetter areas, such as the eastern parts of the country, water storage levels stay mostly the same. This is because any extra moisture is quickly soaked up and stored in the ground.
- In drier, semiarid areas, like the southwest and south-central regions, there is a noticeable decrease in water storage. This happens because these areas often face droughts, which reduce the amount of water stored in the ground.
- In northwestern and north-central US, TWS depletion is significantly reduced, indicating the positive effects of efficient groundwater management and water conservation practices.

The research emphasizes the importance of conjunctive use (the combination of surface water and groundwater) for sustainable water management. By using inefficient surface-water irrigation to promote groundwater recharge and improving groundwater irrigation methods, water management can be optimized. Additionally, managing aquifer recharge plays a key role in mitigating the depletion of vital water sources in regions facing long-term drought.

Salton Sea Management Program

I participated in and made a full submittal to this program. The "independent counsel" that reviewed all of the proposals was composed almost entirely of academics from the University of San Diego. In short, they all had substantial conflicts of interest, and their clear goal was to create "a smaller and more sustainable" Salton Sea with far less evaporation. Like the Tulare and Owens Valley lakes, they intend to make small environmental gestures to appease environmentalists outside of the affected region. They are intentionally drying out another large body of water in CA without concern for the local and regional environmental impacts. As the "playa" or lake bottom dries out, local air quality will continue to deteriorate, and more cases of things like lung cancer, asthma, and other lung-related issues will continue to rise. Drying out all of these regional bodies of water does have a "climate change" impact on the regional water cycle and will continue to decrease annual water flows in the Colorado River.

The Salton Sea Management Program (SSMP) addresses the shrinking of the Salton Sea, which continues through dust suppression and aquatic habitat restoration. These projects aim at safeguarding air quality and rejuvenating wildlife habitats with the objective of restoring a combined 29,800 acres[42] across the sea. These projects have the intention of mitigating the impacts of the disappearing Salton Sea, which offers multiple advantages to the environment. They contribute to making the air cleaner and offering new habitats for migratory birds and aquatic species. A significant undertaking in this regard is the Species Conservation Habitat project, aimed at developing essential water habitats. The development of East Pond 1 is a significant component of this project and is scheduled to be completed by summer 2025. This expansion is crucial for developing fish habitats and decreasing dust from the uncovered lakebeds. All these actions contribute to better air quality and a healthier environment.

The SSMP team is improving vegetation at several locations, such as Clubhouse, Tule Wash, and West Bombay Beach. They are busy planting seeds and adding new plants in these areas. Throughout the year, more plants will grow in the exposed lakebed. This effort is crucial for controlling dust and improving local ecosystems. All these actions follow the guidelines of the Dust Suppression Action Plan.

Salton Sea Conservancy – A Major Milestone!

The formation of the Salton Sea Conservancy will be a key turning point in 2025. This new conservancy was authorized by Senate Bill 583, passed in 2024, and will be the first new state conservancy in 15 years. The Salton Sea Conservancy will oversee the operation and maintenance of completed SSMP projects, ensuring their long-term success and sustaining environmental benefits for the region.

Governor Newsom's January 2025 budget proposal included $147.6 million[43] for Salton Sea restoration projects. This funding was made possible by the Climate Bond, which California voters passed in November 2024. If approved by the state legislature later in 2025, these funds will be

[42] Salton Sea. (n.d.). Retrieved April 30, 2025, from https://en.wikipedia.org/wiki/Salton_Sea

[43] January 2025 - Salton Sea Management Program. (n.d.). Salton Sea Management Program. https://saltonsea.ca.gov/2025/01/

used to accelerate the SSMP's high-priority projects, providing substantial environmental and public health benefits to the region.

This whole thing was a waste of time, money, and resources.

The 14th Amendment and California's Policies

On July 9, 1868, the US federal government passed the 14th Amendment to the US Constitution, granting every citizen the right to due process and equal protection under the law. This guarantees individuals the right to life, liberty, and property, but it raises questions about how California's policies, especially concerning water usage, housing, and environmental issues, might be infringing upon these rights for residents in cities like Los Angeles.

The Impact of California's Water Policies on Citizens

California's water policies have left many citizens vulnerable to extreme conditions like fires and droughts. For example, LA residents face the risk of losing their homes and loved ones due to these policies. When insurance policies are canceled because of government regulations regarding water management, the victims often find themselves at the mercy of federal disaster relief. Rebuilding in high-risk zones raises questions about safety and responsibility. The cost of insurance, driven up by the state's failure to manage these environmental issues, affects homeowners throughout the country.

Rebuilding on High-Risk Land: Is It Worth It?

California's coastal areas, especially Southern California, face a unique combination of wildfires, earthquakes, and landslides, putting residents at considerable risk. Many of these areas are not only prone to natural disasters but also face water scarcity. Rebuilding in these zones, especially on public lands in Owens Valley or regions like Bishop, poses significant challenges related to water availability and environmental sustainability. Should these high-risk areas continue to be developed, or should we consider alternative locations for housing, such as federal lands with access to water resources?

Governor Newsom just asked for $40 billion in federal aid to help with the $250 billion cost of the LA fires[44].

California's Water Infrastructure and the Cost to Taxpayers

The water infrastructure that supports Southern California cities, including the LA Aqueduct, was largely funded by taxpayers at both the state and federal levels. The cost of moving vast amounts of water from far-off sources comes with a hefty price tag, including a $1.19 billion annual power bill to lift 2.5 million acre-feet of water. This expense is passed on to taxpayers and ratepayers across the nation. Meanwhile, areas where this water is sourced, like the Owens Valley, have faced land and water rights disputes, contributing to the "California Water Wars" and environmental damage to the source regions (covered in Chapter 2).

One potential solution to California's water issues is large-scale desalination. If desalination plants could be implemented along the coast, particularly near the Port of Long Beach or more similar locations, they could provide substantial amounts of freshwater. For example, a desalination plant capable of producing 300,000 acre-feet of water per year could support a city of 2.5 million residents. Coupled with efficient water recycling, this could help alleviate some of the strain on California's water systems, providing an alternative to the overreliance on aqueducts from distant sources.

Building a Water-Efficient, High-Density City in Owens Valley

Another proposal to address the housing crisis is the creation of a water-efficient, high-density city north of Bishop, CA, on federal lands. This city would be supported by both local aquifers and a desalination plant. It could house 2.5 million people with a yearly water demand of around 300,000 acre-feet. With advanced water reclamation and conservation systems, this "Freedom City" could offer an innovative solution to California's growing housing shortage while minimizing its impact on the environment.

---

[44] California Governor Requests $40 Billion For Fire Aid, Washington Post Reports. (2025, February 22). Reuters. https://www.reuters.com/world/us/california-governor-requests-40-billion-fire-aid-washington-post-reports-2025-02-22/

The Potential of Data Centers and Water Redundancy

The growing demand for data centers presents another analogy to the challenges California faces in managing resources like water. Data centers require redundant power supplies, and similarly, California's cities need a redundant water supply to mitigate risks from earthquakes, fires, and droughts. Just as data centers rely on backup systems to ensure continuous operation, California must consider how it can ensure a reliable water supply to protect its residents.

Cost of Water in California

The cost of building most of California's water distribution system has long been paid for by the taxpayers and ratepayers. Today, the entire system is owned by the public sector in California. Operating costs consist of maintenance, power, and staff. Moving water is the most power-intensive industry in the state, and California will never give up its natural gas and oil-based power plants that help move the water. More expensive and unreliable forms of power like wind and solar are for the "presents" that exist under the thumb of these out-of-control bureaucracies.

Water costs are a major concern for many people in California. Prices depend on location and usage, and they are usually higher than in other states. This is due to issues like water scarcity, the need to fix old infrastructure, and strict environmental rules. Costs are rising because we need to update and maintain complex systems that deliver water to millions of residents. This involves meeting new treatment standards, finding new water sources, and having skilled workers.

Water bills are increasing faster than inflation, meaning they're rising more quickly than other prices. Sewer charges are going up too, as both water and sewer systems need big investments to ensure they work well and follow regulations. While these higher costs are essential for future reliability, they can be tough on many families, especially those with lower incomes.

Water Affordability Challenges

Many people in California are concerned about how expensive water is. A recent study by the State Water Board found that 21% of the water systems have high prices. These systems provide water to 18% of the residents, and the costs are too high for essential activities like drinking, cooking, and washing. This makes it hard for these people to afford the water they need everyday.

As California's water costs rise, more households may find themselves in a difficult position where they must choose between paying for water and meeting other essential needs like housing, food, and healthcare. This is the inevitable result of overbuilding in a desert environment where water has to be imported through a system entirely owned by socialist bureaucracies. To paraphrase Margaret Thatcher, socialism is great until you run out of other people's money...... or water.

Furthermore, the state has identified that 34% of households earning less than 200%[45] of the federal poverty level may require assistance to afford their water bills. To address these affordability concerns, California is exploring more precise local measures that incorporate additional factors, such as housing costs, to better inform rate assistance programs. These efforts will make sure that even those with less money have access to water. By doing this, we can make life better and easier for everyone in the community.

Basic Water Use and Rate Structures

The amount of water needed for basic human survival—roughly 50 gallons per person per day—is minimal compared to the average household's overall water consumption. Water bills are usually made up of two main parts. The first part is a fixed monthly fee that stays the same no matter how much water you use. The second part is a charge that increases based on your water usage. The fixed fee is important for water companies because it helps them keep their finances steady. This is because their costs don't change much, whether people use a lot of water or just a little. However, this way of charging can be hard for families with low income. They might struggle to afford their water bills because the fixed fee takes up a large portion of the bill, even when they only use a small amount of water for basic needs.

---

[45] New Study Reveals Many American Families Struggle to Afford Basic Personal Care Items and Household Goods. (2013, September 19). Feeding America. https://www.feedingamerica.org/about-us/press-room/new-study-reveals-many-american-families-struggle-to-afford-basic-personal-care-items-and-household-goods

To make water more affordable, agencies can work on controlling water system costs. They can achieve this by improving efficiency, cutting waste, and using new technologies. These actions help keep prices stable and make sure everyone gets the water they need.

Water Prices Soaring in California's Central Valley

As the severe drought continues to worsen in the Western United States, water prices in California's Central Valley have risen sharply. This region, responsible for providing roughly 25% of America's food supply, is facing a crisis like never before. One company under the spotlight is Wonderful, owned by billionaires Stewart and Lynda Resnick. They possess considerable control over a large portion of the valley's water, mainly because of long-term agreements and water rights linked to their vast agricultural properties. Wonderful possesses over 175,000 acres of agricultural land, with almost 130,000 acres situated in California. They consume nearly 150 billion gallons of water annually, which could meet the needs of San Francisco for a decade[46].

The Resnicks acquire roughly 9% of their water from the open market, as they can invest more than many other farmers, giving them considerable sway over local water prices. This scenario has raised worries that the Resnicks could have excessive control over the area's water supplies, particularly in periods of drought. The company contends that their emphasis is on obtaining water rights through land purchases and they lack the ability to directly affect market prices.

Kern Water Bank: A Controversial Resource

At the core of the Resnicks' water activities is the Kern Water Bank, a vast subterranean facility spanning 32 square miles. It has the capacity to hold 1.5 million acre-feet of water[47], equivalent to roughly 500 billion gallons. The Resnicks hold 57% of this bank, co-owning it with various public and private entities. This reservoir enables them to conserve water during arid times, providing them a considerable edge over nearby farmers lacking such facilities.

[46] Sorvino, C. (2021, September 20). Amid Drought, Billionaires Control A Critical California Water Bank. Forbes. https://www.forbes.com/sites/chloesorvino/2021/09/20/amid-drought-billionaires-control-a-critical-california-water-bank/

[47] Kern Water Bank Authority. (n.d.). Kern Water Bank Authority. https://www.kwb.org/

(LA and Friant Kern Canal)

This arrangement has ignited debate. Critics assert that the Resnicks are utilizing both public and private resources to obtain their water supply. Char Miller, an environmental analyst at Pomona College, emphasizes that the Resnicks have shielded themselves from drought. In adjacent regions, underprivileged communities, such as farmworkers, face challenges with water availability and frequently rely on trucks for transport. This scenario illustrates the broader social disparities linked to California's water crisis.

Rising Water Prices Amid Drought

In California's Central Valley, water prices have skyrocketed to $2,000 per acre-foot[48], a huge increase from normal prices of around $250 per acre-foot. Other farmers have sold their future water rights for up to $5,000 per acre-foot in times of extreme drought. With its value depending on the price of the water market, the Kern Water Bank has become a treasure in this unstable market. With the Resnicks' controlling interest estimated between $200 million and $1.7 billion, the bank's present valuation might range from $375 million to $3 billion.

Consequences of Nonpayment and Support Systems

Even though most water companies are dedicated to helping their clients who are struggling financially, nonpayment-related service interruptions may nonetheless happen as a last resort. Utilities often help consumers who are having trouble paying their bills and work with them on repayment options. Shutoffs can be needed nevertheless, in order to guarantee that utilities can recoup the money required to keep services operational.

During the COVID-19 pandemic, a statewide moratorium on water shutoffs was implemented to protect vulnerable residents who were experiencing financial difficulties. While this policy helped ensure access to water during an unprecedented public health crisis, it also resulted in a significant accumulation of unpaid water bills.

By early 2021, more than 155,000 households owed over $1,000 in water bills[49]. In response, the federal government and the state of California have proposed funding initiatives to help reduce this water debt.

As the state continues to navigate these challenges, better data on which customers are struggling to pay their bills will be essential in creating more targeted solutions. Tailored approaches, such as reducing fees on water debt, avoiding shutoffs during emergencies, and

---

[48] Henry, L. (2022, April 26). Third year of drought pushes price of water to $2,000 an acre foot. SJV Water. https://sjvwater.org/third-year-of-drought-pushes-price-of-water-to-2000-an-acre-foot/

[49] Chappelle, C., & Hanak, E. (2021, May). Water Affordability in California. Public Policy Institute of California. https://www.ppic.org/publication/water-affordability/

providing a minimum amount of water to those in need, could help ensure that all Californians have access to this essential resource.

Chapter 4

The Hydrant Crisis: Understanding the Science and Systems Behind It

California's coastal cities, particularly those in Southern California, face a complex set of natural risks that challenge the stability and reliability of their infrastructure, most notably in the areas of fire protection and water supply. As populations in these areas grow (along with the magnitude of the disaster costs in dollars and lives) and the frequency of disasters like earthquakes and wildfires increases, the systems in place to mitigate such risks come under immense strain. This is one of the dangers of urban sprawl in this area. When it comes to disaster preparedness, California's infrastructure particularly its water delivery systems often face the most difficult test since natural threats are both frequent and devastating.

The Intersection of Natural Risks: Fires, Earthquakes, and Landslides

With its bustling coastal cities and expansive wilderness, Southern California is situated at the intersection point of multiple natural disaster zones. Regular landslides, earthquakes, and wildfires in the area provide an unusual combination of hazards that imperil both property as well as essential services. Because earthquakes and wildfires can destroy property and inflict irreversible damage to infrastructure, the insurance business in particular faces a high risk of loss.

The cost of maintaining adequate insurance for properties in such high-risk areas has led to increased premiums for homeowners. In essence, people living outside these high-risk zones, who are largely unaffected by these natural disasters, may bear the financial brunt of the risks faced by those in California.

Southern California Earthquake Probabilities

Southern California is constantly at risk of a catastrophic earthquake, especially in the Greater Los Angeles Area, which encompasses the counties of Ventura, Orange, Riverside, San Bernardino, and Los Angeles. The San Andreas Fault, which runs the length of the state and cuts through Los Angeles County, is the most well-known fault in the area. The San Andreas Fault remains a key area of earthquake risk evaluations for Southern California because of its ability to generate powerful earthquakes. This fault is vulnerable to quakes reaching magnitudes of 8, potentially leading to significant devastation and disrupting the water systems essential for the area's survival.

Earthquakes pose a risk beyond Southern California. San Francisco faces a high risk due to its location on the Pacific Ring of Fire. This area is famous for its frequent and powerful earthquakes. The city sits near both the San Andreas Fault and the Hayward Fault, major tectonic boundaries where the Pacific and North American plates meet. The movement of these plates causes stress to build up, which is eventually released as earthquakes. Additionally, certain areas in San Francisco are built on soft soils or reclaimed land, making them highly susceptible to liquefaction during strong shaking. Liquefaction occurs when saturated soils lose their strength and behave like a liquid, leading to severe damage to buildings and infrastructure.

Historically, the Southern San Andreas Fault has seen significant seismic activity, but the timeline between major events is far from predictable. The last major earthquake on the fault occurred in 1857, while another was recorded as far back as 1812. The approximate 45-year intervals between these earthquakes do not provide a consistent pattern, leaving uncertainties around when the next major event may occur. The San Andreas Fault does not operate on a regular schedule, and as such, it is difficult to predict the exact timing of its next major rupture.

Paleoseismology is all about studying old earthquakes by looking at clues in rocks and soil. This study teaches us about how faults move and cause shaking. In the area from Palm Springs to the Salton Sea, big earthquakes don't occur very often, generally every 200 to 300 years. The last major earthquake happened roughly 300 years ago when Spanish explorers were around. Unfortunately, there are no written records of that event. Scientists use tools like radiocarbon dating to find signs of these earthquakes. These methods have revealed evidence of several big earthquakes going back to 800 A.D. This helps us to have a clearer understanding of the earthquake history in the region. With a 3,000-year record, Wrightwood, California, is one of the most well-documented paleoseismology sites. The average time between massive earthquakes is approximately 100 years, according to statistics from this site that combines data from several research. This record implies that big earthquakes on the southern San Andreas Fault are not sporadic but rather follow a fairly regular cycle, even though exact date is unknown. Since the 1857 event, the current time interval between two large earthquakes has only been surpassed four times in history. The danger for Southern California is significantly heightened by the news, which suggests that a large earthquake on this section of the fault is likely to occur soon.

Magnitude 6.7 or Higher: The Odds of a Major Earthquake

The US Geological Survey (USGS) regularly refreshes its predictions to estimate the likelihood of significant earthquakes. As per their most recent report[50], there's a 99% likelihood that a magnitude 6.7 or greater earthquake will strike Southern California in the coming 30 years. This significant likelihood underscores the critical necessity for readiness. Preparation is crucial since a major earthquake can greatly harm vital infrastructure, such as water systems, leading to substantial interruptions in everyday life.

Although the San Andreas Fault is the most recognized and talked about fault in Southern California, it is certainly not the sole source of seismic danger. The 1994 Northridge Earthquake served as a clear warning about the threats presented by obscure faults. The Northridge

[50] Forecasting California's Earthquakes—What Can We Expect in the Next 30 Years? (2008). In USGC. https://www.usgs.gov/index.php/publications/probability-one-or-more-m-7-earthquakes-southern-california-30-years

earthquake, stemming from the Northridge Blind Thrust Fault, caused 57 fatalities, more than 9,000 injuries, and economic damages exceeding $49 billion[51]. The earthquake also activated movement in several additional fault zones, indicating that seismic hazards in Southern California are much more intricate than just concentrating on the San Andreas Fault.

In 2019, the Ridgecrest Earthquakes rattled California, representing the biggest seismic occurrence in the state in over two decades. These tremors, which hit close to the Mojave Desert and impacted several fault lines, such as the Owens Valley, Panamint Valley, Garlock, and Blackwater Faults, showed that numerous seismic events can happen at once, increasing the dangers faced by the area. The earthquake on July 4th, which registered a magnitude of 6.4, was succeeded by a more powerful 7.1 magnitude quake the following day, highlighting the possibility of several aftershocks amplifying the destruction from an already major incident.

Escalating Earthquake Risks and Their Possible Impact on Water Infrastructure

The Ridgecrest earthquakes temporarily increased the likelihood of a major earthquake on California's San Andreas Fault—a condition that existed for several months after the quake. The probability of a magnitude 7.5 or greater earthquake on the Garlock Fault—a fault running along the northern edge of the Mojave Desert—rose significantly. This, in turn, slightly elevated the chances of a major rupture on the San Andreas Fault, which comes within 35 miles of downtown Los Angeles. However, the long-term and more pressing risk of a major quake on the southern portion of the San Andreas is tied to the fact that this segment of the fault has not experienced a significant rupture in approximately 350 years. Seismologists have warned of this overdue rupture for the past 50-60 years, a concern that even influenced Hollywood's release of the movie Earthquake in 1974. While the film portrayed an exaggerated magnitude 9.9 quake, a real seismic event on the southern San Andreas would likely range from 7.9 to 8.3—similar in size to the 1906 San Francisco earthquake.

[51] Earthquake Facts: The Northridge Earthquake: January 17, 1994. (n.d.). Earthquake Country Alliance. https://www.earthquakecountry.org/northridge/facts/

Although the odds of any single earthquake disrupting water service remain low, the sheer number of aqueducts crossing fault lines raises concerns. The 2019 Ridgecrest earthquakes remind us that even distant seismic activity can set off a chain of events, increasing stress on adjacent faults. If the Garlock Fault were to rupture close to the San Andreas, there could be a substantial chance of an immediate or near-future rupture along the San Andreas as well.

One of the most significant concerns following a major quake is the potential for post-earthquake fires, similar to those seen after the 1906 San Francisco earthquake. Damage to aqueducts could hinder firefighting efforts, exacerbating destruction in urban areas. This emphasizes the need for continued investment in improving early warning systems and ensuring that critical resources like water remain accessible in the wake of a major disaster.

Water Infrastructure and the Economic Impact of California's Water Systems

California's water systems have been built over many years to accommodate an increasing population and bolster the economic drivers of Southern California. The main aqueducts, including those supplying Los Angeles, traverse high-risk areas like the San Andreas Fault, thereby increasing the water system's susceptibility. The Los Angeles Aqueduct, specifically, serves as one of the major water sources for the area, moving large quantities of water across many miles. The monetary consequences of transporting water over such landscapes are vast. Although the LA Aqueduct primarily transports water downhill, moving this water remains costly. At the same time, the CA Aqueduct raises 2.5 million acre-feet of water annually[52] uphill, covering an elevation of 3,116 feet from the Delta to the northeastern high desert of LA.

Through calculations grounded in the energy required to elevate water, the yearly power expense for transporting water via the aqueduct system might amount to $1.19 billion. Such an elevated cost, influenced by the considerable elevation variations in California's water supply

[52] California Aqueduct. (n.d.). Water Education Foundation. https://www.watereducation.org/aquapedia/california-aqueduct

systems, highlights the immense infrastructure obstacles faced in supplying water to urban areas in Southern California.

About halfway between the Bay Delta and "the Big Lift' that moves water from the Southern floor of the central valley 1,926 feet uphill into the high desert Northeast of LA lies the only "Southern reservoir" in the California system. This is an "off channel" reservoir where water is pumped uphill, stored for a time then released back into the aqueduct for use at points further south. This type of reservoir is called a "pump-generate" water storage facility as when the water stored in the elevated reservoir most of the energy required to pump it uphill is recovered by a generator.

In wet years in California, pumping from the Delta is halted when the San Luis reservoir is filled. Adding a second off-channel reservoir at the location shown below would be far more cost-effective and productive than raising or building more reservoirs in the north. This concept reflects an older concept proposed for Delta exports, termed the "big gulp little sip" concept. This common-sense concept simply means pump more water when it is available and less when it is not.

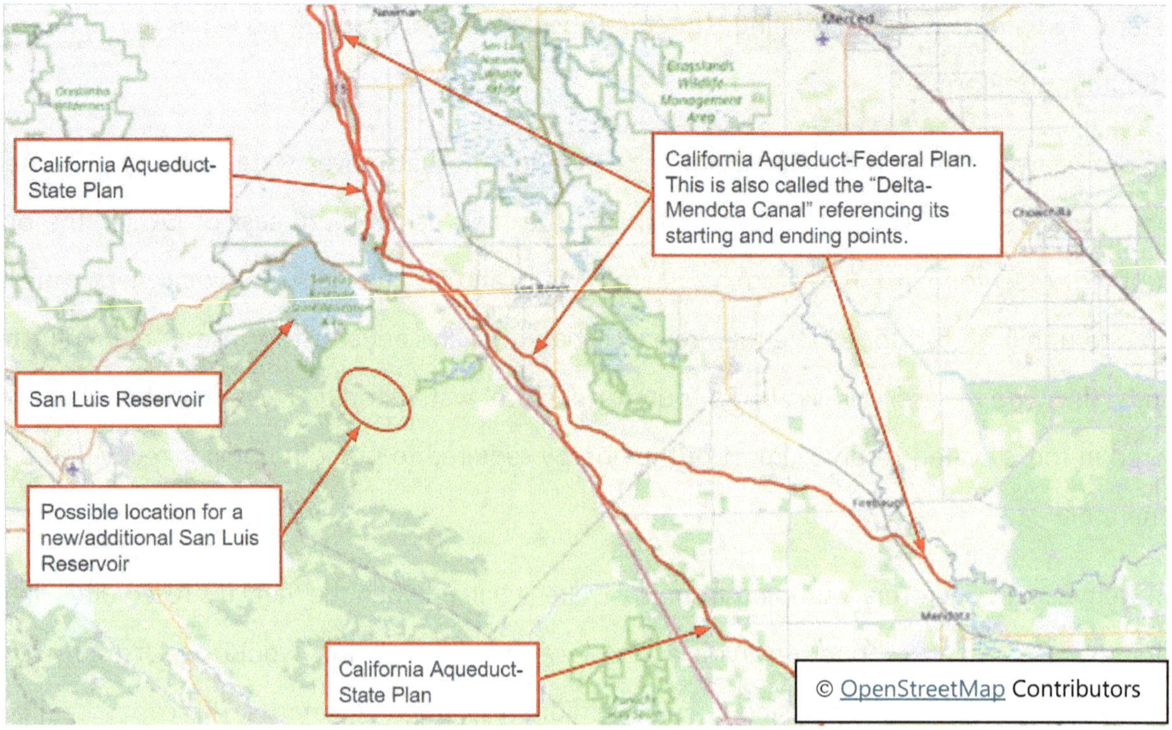

(Delta Mendota Canal)

However, the real danger is that the water supply can be interrupted by a natural calamity, like a large earthquake. The entire California water system has extensive exposure to earthquake risk. A breach is likely since the main aqueducts that supply California's cities overlap fault lines. A major catastrophe might cut off the water supply, resulting in serious harm and disturbance. It could also hinder efforts to put out wildfires, which would make matters worse.

The Need for Redundancy and Resilience in Water Systems

Similar to how data centers need backup power sources to prevent significant data loss during outages, California's water systems also require redundancy to guarantee water availability during emergencies. Water storage in nearby reservoirs acts like a "battery backup" for the water supply, making certain that there is sufficient water for firefighting, drinking, and other critical requirements, even when conventional water delivery methods fail. An extra water source through desalination facilities, which would offer an alternative source of freshwater, can further bolster California's resilience to disasters like wildfires, earthquakes, and droughts.

Constructing desalination facilities along the coastline would provide an extra, consistent source of water that would alleviate the impact of a natural disaster. The plants would enable increased local storage and distribution of water, creating a buffer against disruption of large-scale water systems susceptible to seismic movement. This upgrade in infrastructure is crucial since the current old systems have massive dependence on water transport long distances that can readily be severed amidst disasters.

The Social and Economic Implications of Water and Fire Protection

Beyond the technical challenges of maintaining and expanding water infrastructure, there is also a social and economic dimension to the crisis. As mentioned, the high risk associated with fires, earthquakes, and landslides in California increases the overall cost of living in the state, particularly in areas with high property values. These dynamic places a burden not just on homeowners but also on taxpayers, who indirectly finance the rebuilding and recovery efforts through various state and federal programs.

In this context, the hydrant crisis is not just about physical infrastructure but also about policy decisions and economic systems that influence the ability of communities to effectively prepare for and recover from natural disasters. Whether through better insurance policies, the development of more resilient infrastructure, or a more equitable distribution of resources for disaster response, these systemic challenges must be addressed to ensure the long-term safety and sustainability of California's communities.

The recent fires across the state of California are causing home insurance rates to rise significantly, as well as causing insurance companies to cancel policies in their entirety. It is becoming clear that insurance rates in California have been artificially low relative to the risk of loss for decades.

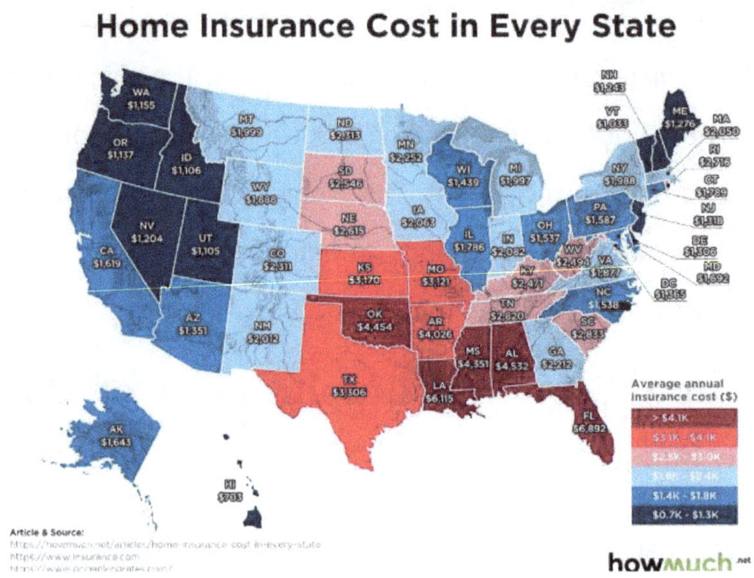

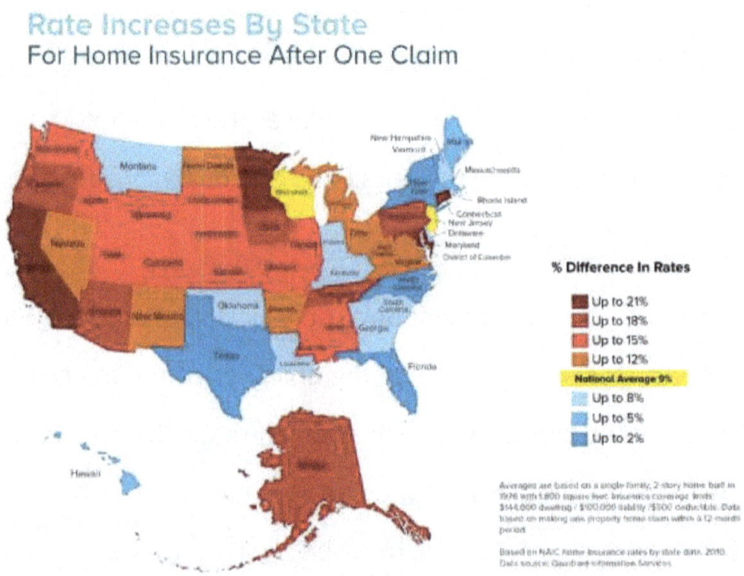

Today, rates are rising to reflect the risks associated with earthquakes, fires, wind events, and landslides.

Water Restoration Plans

For over two decades, California's water policies, such as the "conserve and transfer" strategy, have prioritized urban water use over agricultural and environmental needs at the locations where the water comes from. This has led to the depletion of critical freshwater resources that sustain both ecosystems and agricultural production in the region. As water for agriculture has been increasingly diverted or reduced, the surrounding ecosystems, including the Lower Colorado River Delta, are facing rising salinity levels, reduced water availability, and increasing desertification. These water policies have further intensified tensions between states, as Arizona and neighboring regions now face the consequences of California's growing influence over water management, potentially "dehydrating" the Southwest. With water scarcity becoming a more pressing issue, the restoration of the CRD is critical not only for environmental health but also for regional stability.

Prescott Water Management and Restoration Plan

The fight for water resources in the Prescott area, particularly surrounding the Big Chino and Little Chino aquifers, has escalated over the past decade, with the local population facing two main options: do nothing or invest in an expensive $400 million welded steel pipeline. The proposed pipeline would convey water from the Big Chino aquifer to the Little Chino aquifer for domestic use and economic development, but it raises concerns about negative impacts on the Verde River's riparian environment. While the pipeline solution is one approach, the goal is to explore alternative, more sustainable solutions that offer a balance between economic growth and environmental preservation.

At present, the Verde River receives about 10,000 acre-feet per year[53] of flood flows from brief and intense storm events, which ultimately flow into Horseshoe and Bartlett reservoirs. The US Geological Survey (USGS) streamflow station near Paulden estimates these flows, showing how excess water is often lost to downstream reservoirs. Meanwhile, projects such as the Big Chino Water Ranch propose to pump between 8,000 and 12,000 acre-feet annually from the Big Chino aquifer, further reducing Verde River flow, which is already facing declines.

One such potential alternative is based on the Prado Dam model in California, where wetlands and flood control measures are integrated to support aquifer recharge and ecosystem restoration. By creating or rehabilitating wetlands above and below Sullivan Dam on the Upper Verde River, it is possible to restore base flows to a minimum of 20 cubic feet per second (CFS), which is vital for maintaining the river's riparian environment. The proposed solution would involve capturing

---

[53] Haney, A. J., Turner, S. D., & Springer, E. A. (et al.) (2008). Ecological Implications of Verde River Flows. In Arizona Conservation Science. https://azconservation.org/dl/TNCAZ_VerdeRiver_Ecological_Flows.pdf

floodwaters, similar to the flood control and recharge systems in Orange County, where Prado Dam provides a mechanism for impounding floodwaters and releasing them gradually to support aquifer recharge. Capturing around 5,000-6,000 acre-feet (AF)[54] of floodwater on a periodic basis could significantly reduce the dependency on water from the Big Chino Water Ranch by 2,000 AF per year, decreasing the need to pump groundwater.

This strategy would not only help protect water quality and quantity in downstream areas like Clarksdale but also address the decreasing water table in the Little Chino aquifer, which has been dropping for decades. Additionally, the restoration of wetlands and riparian environments could provide long-term benefits for endangered species, similar to the success seen in the Prado Wetlands. While this proposed solution is not perfect, it offers a compromise that balances environmental sustainability with the water needs of the local community, ultimately allowing the Prescott area to move beyond the binary choice of doing nothing or investing in a costly pipeline.

The plan further includes creating a linear city along transportation and water corridors, which would integrate high-density residential housing and a robust water reclamation and recharge system. This model aims to reclaim up to 80% of residential water use and stormwater, contributing to the aquifer recharge and reducing reliance on imported water. The project also aims to implement advanced water reclamation systems (AWRS) in the Prescott and Prescott Valley areas, improving water efficiency and reducing the need for new water sources.

The project's goal is to infiltrate approximately 6,000 acre-feet of stormwater over a 45-day period, with the aquifer serving as the primary form of water storage. In addition to stormwater capture, the design includes a road-based utility system for stormwater collection and wastewater treatment, leveraging reverse osmosis (RO) technology for water reclamation.

Innovative Water Management

California can explore innovative methods of water management and restoration by harnessing evaporation and wind patterns. The Tres Meres and Prescott Water Management and Restoration

[54] Cook, D. (2007). Chino Valley Approves Pipeline. Center for Biological Diversity. https://www.biologicaldiversity.org/news/media-archive/BigChino5-12-07.pdf

Plan can benefit from the Danakil model by creating flood channels or using natural topography to redirect water into strategic areas for evaporation and condensation, ultimately boosting local water supplies. Furthermore, by focusing on the *Freshwater-Food-Energy-Carbon Nexus*, these plans could integrate renewable energy generation and climate-friendly irrigation systems, mirroring the Danakil approach's synergy between water, energy, and environmental restoration.

Chapter 5

A Proposed First Freedom City Framework

A proposed site for a new freedom city is located approximately 90 miles from the Grand Canyon on federally managed land. This site offers several logistical and geographic advantages that align with national development priorities. It is positioned at an elevation of 4,650 feet, providing a favorable climate for year-round living and tourism.

The location has access to critical infrastructure, including:

- Hydroelectric power transmission lines
- Two existing natural pipelines
- Rail connectivity
- Developed road access
- Local production capacity for concrete, steel, and lumber

However, one major infrastructure gap is the absence of a regional airport capable of landing aircraft such as the Boeing 737. This issue would need to be addressed to support tourism and economic growth. For context, Phoenix Sky Harbor Airport is located 221 miles from the Grand Canyon, and Las Vegas Airport is 274 miles away. A new airport in this proposed region would provide better proximity to the Grand Canyon and improve transportation access.

Economic Potential and Public Land Access

Tourists currently spend approximately $759 million annually[55] visiting the Grand Canyon. Despite this, large sections of nearby public lands are restricted from public use. This current land management model has been criticized for benefiting a select few while denying the general public access to lands they collectively own. Reassessment of these policies may open opportunities for broader public benefit.

The Big Chino Water Ranch (BCWR) Project

The Big Chino Water Ranch outlines a major water infrastructure project targeting the Big Chino aquifer. Originally developed decades ago, the project was intended to supply water to the rapidly growing Prescott region. The City of Prescott approved $161,000 in engineering contracts[56] to update cost and feasibility estimates. The overall project is estimated to cost between $170 million and $174.8 million, including land acquisition, construction, rights of way, permitting, and engineering.

Criticism of the BCWR project centers on its high cost and environmental impact. The project would divert approximately 14,000 acre-feet of water per year from the headwaters of the Verde River. Minimal, if any, of that water would be returned to the river, impacting downstream users and regional ecosystems. The project has been characterized as a mechanism to subsidize luxury housing for higher-income residents using external water sources.

In contrast, a smaller-scale, high-density freedom city could house approximately 50,000 residents while consuming less than 1,200 acre-feet of water annually—significantly reducing environmental impact and downstream water loss.

---

[55] Tourism to Grand Canyon National Park Contributed $759 Million To Local Economy in 2022. (2023, August 21). [Press release]. https://www.nps.gov/grca/learn/news/2022-visitor-spending-to-grand-canyon-national-park.htm

[56] Giddens, J. K. (2023, November 22). Big Chino Pipeline Still Probably a Decade Out. Sedona Red Rock News. https://www.redrocknews.com/2023/11/22/big-chino-pipeline-still-probably-a-decade-out/

Water Overuse in Prescott and Surrounding Areas

Prescott and Prescott Valley have continued housing development despite ongoing water shortages. The Little Chino Aquifer is currently in a state of perpetual overdraft. As a result, residents in nearby Chino Valley must drill approximately 70 feet deeper to access groundwater. Some communities are now forced to haul water to their homes due to insufficient well output. Despite these indicators of water stress, construction of new housing tracts continues at pace.

This pattern of development raises long-term concerns about sustainability and risk management. The area is also highly susceptible to wildfires. The region was the setting for the movie Only the Brave, which depicted the Granite Mountain Hotshots—19 firefighters who died in a wildfire near Prescott. Unchecked development in a fire-prone region without resolving water infrastructure issues increases the likelihood of future disasters.

Prescott and Prescott Valley may have already exceeded their sustainable growth limits. Unless development policies are adjusted to align with environmental and infrastructure constraints, continued expansion may result in increased risk and long-term instability.

In contrast to traditional housing models, a proposed high-density, small-scale freedom city could maintain similar indoor water use (approximately 147 GPD per household) while reducing outdoor use to nearly zero. This would allow for the construction of more homes using significantly less water per unit. Furthermore, reducing outdoor water usage would minimize the amount of piping needed to supply and remove water from each home.

With indoor use accounting for all water delivered, the system could route 100% of wastewater to an advanced water reclamation facility. These facilities are capable of reclaiming 90–95% of household water to a high standard, suitable for environmental applications such as restoring flow to the Verde River, though not yet safe for drinking.

Legal Dispute Over Big Chino Water

The City of Prescott's plans to draw groundwater from the Big Chino Sub-basin have met resistance from Phoenix's largest water provider, Salt River Project (SRP). SRP argues that Prescott has no legal entitlement to the water under a 1991 state law and has formally challenged the city's claim in correspondence with both Prescott and the Arizona Department of Water Resources (ADWR).

Prescott maintains that it is entitled to 9,571 acre-feet (af) of groundwater annually. The ADWR's preliminary conclusion estimates this figure at 8,717 af per year. The Salt River Project (SRP) contests both figures, asserting a much lower estimate of 4,081 af annually. The disagreement stems from differing interpretations of Arizona Revised Statutes §45-555(E), which outlines the legal framework for groundwater transfers.

Prescott officials, including Big Chino Water Ranch Manager Jim Holt and Prescott Valley Mayor Harvey Skoog, reject SRP's position, stating that ultimate authority rests with ADWR. Yavapai County Supervisor Carol Springer, who co-authored the original 1991 law—criticized SRP's argument as unfounded and clarified that the law did not create new rights for Prescott but rather affirmed existing ones.

The 1991 legislation prohibits inter-basin groundwater transfers, with these exceptions:

1. Three exceptions for Phoenix-area water providers.
2. One specific exception for Prescott, allowing up to 14,000 af per year from the Big Chino aquifer.
3. An additional provision permitting Prescott AMA municipalities to convert retired irrigation rights to municipal use.

Historical Context and Environmental Concerns

The legal provision for Prescott originated from negotiations related to the Fort McDowell Yavapai Nation's water rights settlement. At one time, planners considered transporting water from the Big Chino aquifer to Fort McDowell via the Verde River. However, environmental concerns from the SRP and U.S. Fish and Wildlife Service halted the plan. A Bureau of Reclamation study suggested that modifying Verde River flow could negatively impact native fish species, dubbed the "Goldilocks fish", due to their sensitivity to water level fluctuations.

Although the Fort McDowell settlement was ultimately resolved through other means, the special exemption for Prescott remained in law.

A Modern Approach to Development

Counties such as Yavapai and Coconino lack the financial capacity to fund large-scale projects independently. However, regional investments in tourism infrastructure—such as a new airport serving visitors from the East Coast and other markets—could generate the necessary local revenue to fund water-related projects.

Costs for infrastructure such as the BCWR include not only engineering and permitting but also materials like steel piping and the labor-intensive tasks of trenching and welding for installations involving three-foot diameter pipes buried with a minimum of two feet of cover.

As Arizona and other western states advance new groundwater management and water recycling regulations, modern and more efficient models of development are increasingly necessary. With thoughtful planning and the support of both local and federal entities, Northern Arizona could become a model for sustainable, water-efficient growth.

Water Infrastructure and Management

The aquifer below the proposed site provides a reliable source of groundwater. With a dispersed well placement strategy, a short network of PVC pipelines would be sufficient to connect municipal wells—minimizing both construction costs and environmental impact. Most of these wells would be located on federal land.

Water sustainability is a cornerstone of this city's design. Approximately 14,000 to 18,000 acre-feet of water currently percolates annually from the Big Chino Aquifer into the Verde River. The proposed plan would limit groundwater withdrawals to approximately 7,000 acre-feet per year and return about 5,800 acre-feet per year of high-quality reclaimed water—classified as Grade A+ effluent—back to the Upper Verde River.

With modern wastewater treatment, over 90% of indoor water use can be reclaimed and repurposed. This reclaimed water could either support environmental restoration or be used to irrigate local amenities such as tree-lined streets, parks, and school grounds. However, the more water diverted for irrigation, the less is available for natural systems and downstream communities.

Urban Design and Population Capacity

The proposed city would span approximately 2.25 square miles, accommodating between 40,000 and 50,000 residents through high-density housing. The compact design reduces the demand for long-distance water transport and sewage lines, further lowering infrastructure costs.

Key infrastructure components include:

- A water main.
- A sewage main line.
- Connections to existing utilities.
- A wastewater reclamation facility with return flow to the Upper Verde River.
- Approximately 40 miles of paving existing roads.
- A 600-acre industrial park surrounding an airport site.

Supplying water to and reclaiming water from any new cities built within the arid region will be a priority. Using the "waste" or "process heat" from a smaller-scale local power plant would be an efficient means of heating water for the city and would be an additional and optional benefit. Even the waste heat from a local Data Center could be used to preheat the water supply for the city.

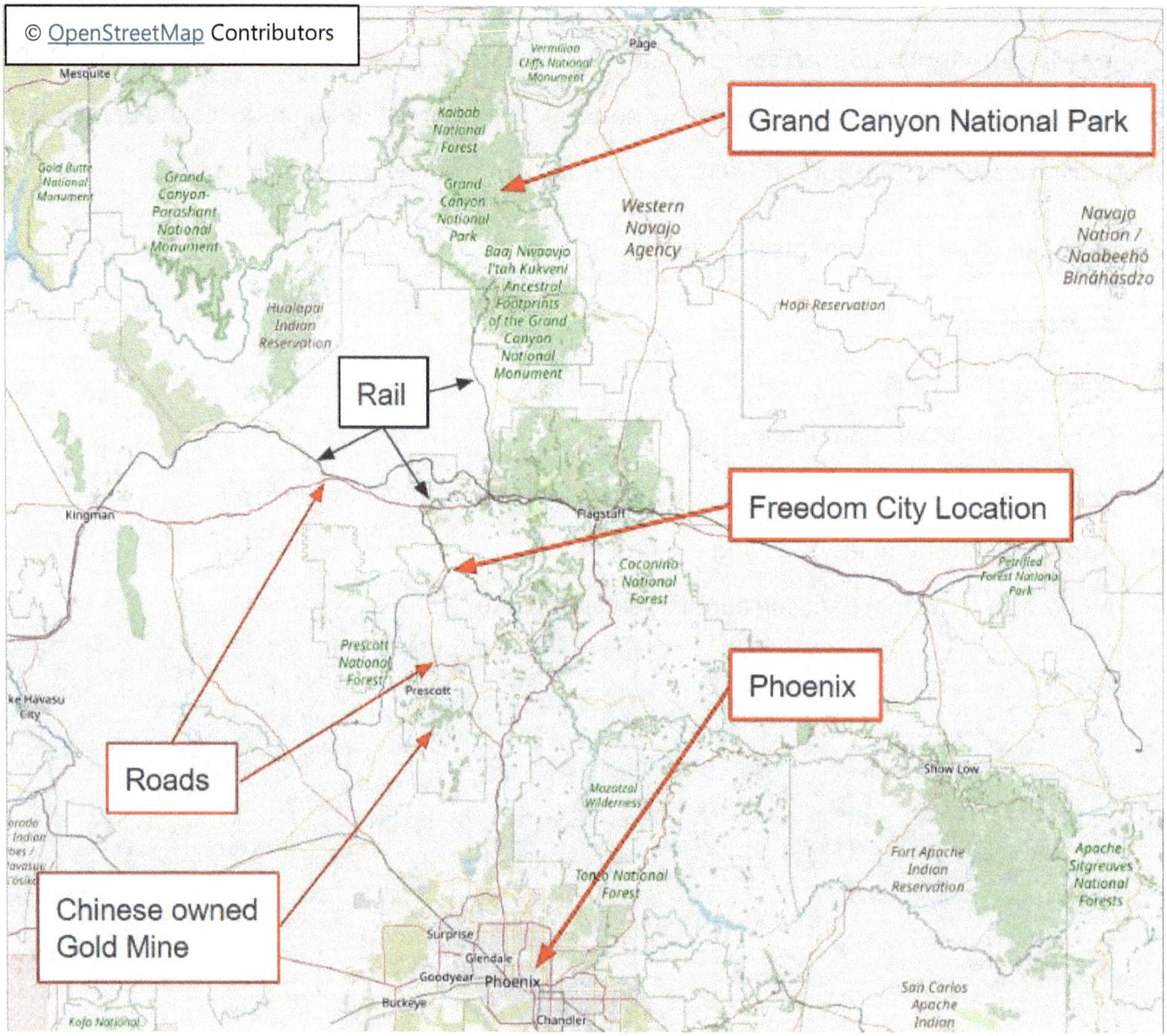

(Site Location Map Image)

The location also presents unique transportation advantages. small airport capable of servicing plains like the 737 or larger is proposed to be built on Wagon Tire Flat, 3 miles North of the Freedom City location shown above. The airport is also located within the Prescott National Forest boundaries. A small airport is proposed to be built on Wagon Tire Flat, located within the National

Forest boundaries of Yavapai County. This area offers sufficient space for the airstrip and associated facilities.

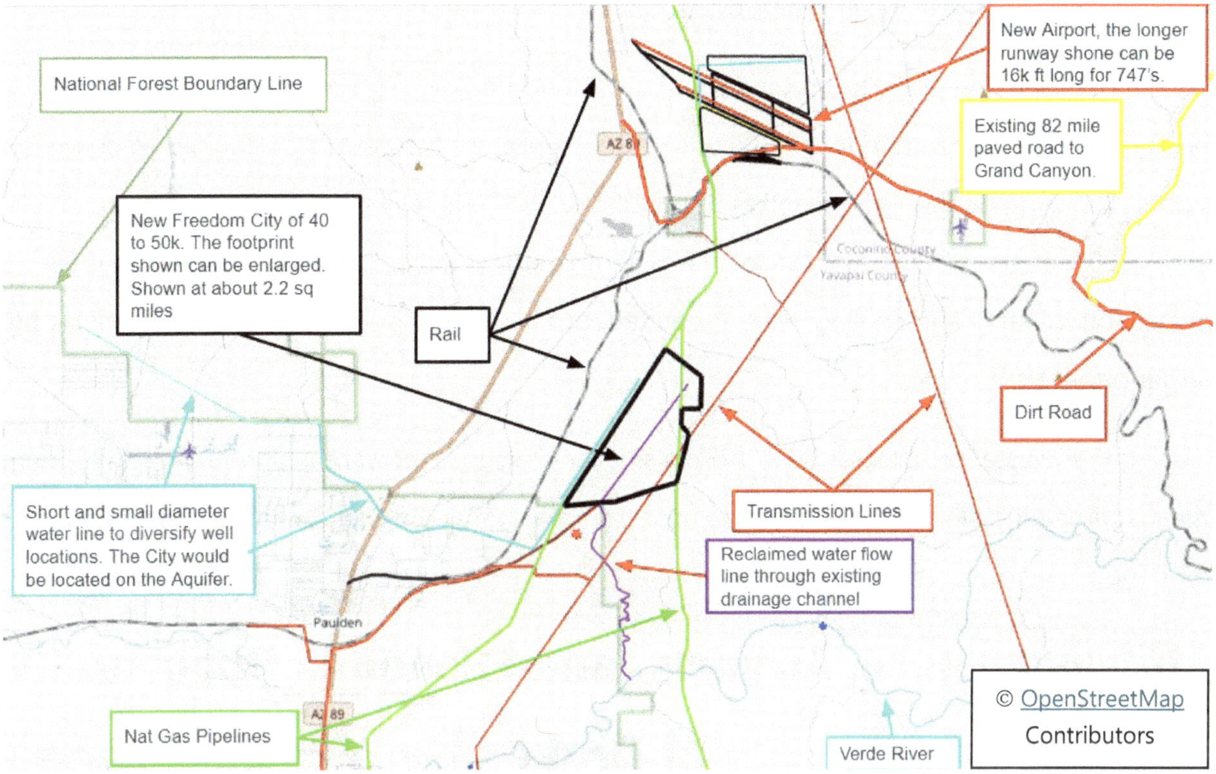

(Site Detail Map Image)

Paving approximately 10 miles of Forest Service Road 71 would connect the city to 81.8 miles of existing paved roads leading directly to the Grand Canyon Village—offering visitors faster and more efficient access than from Phoenix.

This location would also benefit from proximity to existing rail infrastructure. Provided that operational rail contracts for the Grand Canyon Railroad and Verde Canyon Railroad can be renewed, the site would have immediate access to both freight and tourism rail lines.

This alignment supports a tourism-driven economic model: making it cheaper and faster for visitors to reach Northern Arizona, increasing local revenue while reducing strain on Phoenix-based infrastructure.

The specific design of each Freedom City will be a bit different, with architecture and engineering that reflect the topography, geology, and hydrology of each location. A large number of smaller well-spread-out freedom cities can provide the same total number of new housing units and utilize smaller, more responsive city governments. This concept is easier on the residents, the environment, and downstream water users.

For most locations of Freedom Cities within the Arid Region, high mountain valleys that have been utilized as transportation and utility corridors represent the best opportunities to build new population centers. These new centers of population will generally be at altitudes between 4,500 and 6,000 feet in elevation. At these locations, temperatures will be generally cooler than most US cities. Both heating and cooling will be required, but heating during the cold winters will be the larger priority.

Using the "waste" or "process heat" from a smaller-scale local power plant would be an efficient means of heating water for the city. The heat from the water can then be transferred using heat exchangers and used for other purposes.

Infrastructure, Airports, and Economic Opportunity at Drake

The location near the Drake Cement Plant in Yavapai County represents a rare convergence of critical infrastructure and natural feasibility for city development. On-site and surrounding infrastructure include:

- Existing power transmission lines
- Natural Gas Pipelines
- Operational rail lines
- Paved and unpaved forest service roads
- Close access to Interstate Highway 89

The region also has available water resources, a relatively flat topography, and the existing Drake cement kiln is located on 122.85 acres of privately owned land (zoned RCU-2A). This small section of private land is surrounded by vast public lands, making the site ideal for scalable and responsible expansion.

The current commercial operations in this area are likely "grandfathered in," which creates the potential for future tourism-related projects to be entitled under adjoining land exemptions or in alignment with the National Forest mandate, which includes tourism as a legitimate land use.

An Opportunity to Expand Access to the Grand Canyon

The Grand Canyon National Park welcomed over 4.5 million visitors in 2021[57], according to the National Park Service. These visitors contributed approximately $710 million to local economies, supporting over 9,300 jobs, $324 million[58] in labor income, and generating a total economic output of $945 million in the gateway communities. Yet, despite this enormous demand, access remains limited—primarily funneled through a single point of entry.

This imbalance between demand and access represents a significant economic bottleneck. A secondary airport near Drake—just 91 miles from the Grand Canyon, with 81 miles already paved—offers a practical solution. It would enable tourists from the East Coast and around the world to "spend less time getting there and more time being there."

A Commercial-Scale Airport at Drake

The proposed airport near Drake would feature two runways:

- A long runway (~16,000 linear feet) capable of accommodating 747-sized aircraft for transcontinental and international flights
- A shorter runway suitable for 737-class aircraft, commonly used in domestic commercial aviation

For context, the altitude at the proposed site is around 4,710 feet, only modestly lower than Denver International Airport's 5,325 feet. The runway site ranges from a high point of 4,710 feet to a low of 4,670 feet, resulting in a slope of just 0.25%—well below the federal maximum runway slope of 2% and visually indistinguishable from perfectly flat ground.

---

[57] Tourism to Grand Canyon National Park Contributed $710 Million To Local Economy in 2021. (2022, June 27). [Press release]. https://www.nps.gov/grca/learn/news/visitor-use-spending-to-grand-canyon-2021.htm

[58] Ibid.

This location lies between two large mountain ranges, offering relatively easy air access. It stands in stark contrast to many remote western locations where the terrain significantly limits aviation development.

Funding and Feasibility

Financing the airport and associated road upgrades should involve a public-private partnership (P3) model. Legal frameworks already exist to support such ventures, including mechanisms outlined by Kaplan Kirsch LLP for funding and operating airports through P3s. Moreover, the federal government has made significant resources available for airport infrastructure through the Infrastructure Investment and Jobs Act.

As of 2024, the Federal Aviation Administration (FAA) has allocated $2.89 billion annually to US[59] airports for four consecutive years. To date, nearly $12 billion in Airport Infrastructure Grant (AIG) funding has been distributed nationwide—indicating ample opportunity to secure financial support for a project of this magnitude.

Regional Benefits and Environmental Integration

Drake's infrastructure-rich location offers additional advantages:

- Direct access to existing rail lines, ideal for freight and tourism development
- Efficient linkage to surrounding areas through modest paving projects
- Strategic integration into regional economic plans, including 600-acre industrial zones and public land tourism models

Because the proposed airport and supporting infrastructure are embedded within or adjacent to federal land, the project must prioritize access equity. That is: ensuring that the American people—the rightful stakeholders in this land—benefit through improved access, increased economic activity, and enhanced environmental management.

---

[59] Infrastructure Investment and Jobs Act. (n.d.). Federal Aviation Administration. https://www.faa.gov/iija/airport-infrastructure

Managing Arizona's Land and Water for the People

Wildfires have become one of the greatest recurring threats to western states, exacerbated by decades of poor forest management and climate variability. In response, advanced aerial firefighting capabilities have emerged as a critical tool in suppressing fast-moving blazes.

Leading the charge is 10 Tanker Air Carrier, an aerial firefighting company that has been active since 2006. Starting with a single aircraft, the company now operates four DC-10-30 air tankers: Tanker 910, Tanker 911, Tanker 912, and Tanker 914. These aircraft are wide-body McDonnell Douglas DC-10s retrofitted with exterior belly-mounted tanks capable of carrying up to 9,400 US gallons (45,000 liters) of water or retardant. Each aircraft can discharge its full payload in just eight seconds, making them indispensable in initial attack and extended wildfire suppression efforts across the western U.S.

The Gold Paradise Peak Mine

Unfortunately, misaligned policy has left the door open for foreign exploitation of public lands. In the Bradshaw Mountains, near the community of Potato Patch and just 10 miles south of Prescott, a group of mining companies linked to Chinese investors have launched the development of the Gold Paradise Peak Mine.

The operation spans over 400 acres in private ownership and 1,000 acres of claims in Prescott National Forest, directly threatening Mount Tritle—a prominent local peak—and the Hassayampa River, home to some of Prescott's most cherished waterfalls and swimming holes. Already, the mine faces violations of the Clean Water Act, Endangered Species Act, and state permitting requirements.

No American mining company has pursued this venture, largely because the U.S. regulatory process and compliance costs make it economically nonviable. This imbalance reflects broader policy failures and a tilt toward "Muirist" preservationist ideology over the more balanced and utilitarian Pinchot approach to land management.

The Muir-Pinchot Divide

The debate over American land use policy regarding the split between John Muir (preservation) and Gifford Pinchot is more than philosophical. It affects fire risk, rural economies, water access, and even national security.

Along the Hassayampa and Lynx Creek watersheds, a lack of proactive management has made these areas vulnerable to catastrophic fires and foreign encroachment. The situation highlights the need to rebalance federal forest policy toward a science- and economics-driven strategy—one that supports domestic forestry jobs, controls fire risk, and secures access to natural resources for U.S. citizens.

This isn't a new issue. The same policy debate has echoed for over 150 years, as noted by John Wesley Powell, who once proposed that watershed boundaries should define political regions in the West. He reasoned that to avoid monopolizing water rights, the people who live on the land must also control the water that sustains it.

Sustainable Growth in the Arid West

The arid Southwest faces a persistent and worsening water crisis. Expanding high-density urban areas further into desert terrain will only lead to aquifer overdraft, system inefficiencies, and regional conflict over scarce water.

Rather than pump more water uphill into aging urban cores, we must pivot to modular, high-density housing solutions paired with localized modular water reclamation systems. These "off-the-shelf" technologies could dramatically increase population capacity and economic productivity in counties like Yavapai—without increasing strain on the water supply.

Rapid percolation systems should be built to return reclaimed water into natural aquifers. Cities of the future in the American West must function as closed-loop systems, not open drains. Only by decoupling growth from aquifer overdraft can we avoid the slow collapse of regional water systems.

The Democratic Use of Shared Lands and Waters

As the conversation returns to core democratic principles, one truth becomes clear: the National Forests and Parks of the United States belong to all Americans. These lands should be managed not for political gain or foreign exploitation but to ensure maximum benefit to the greatest number of citizens.

The future of the Colorado River Basin and the greater Arid West lies not in further polarization but in reconciliation between the Muir and Pinchot philosophies. The middle path offers the most democratic and resilient vision for the West.

Chapter 6

Freedom Cities

Across the United States, not just in California, the nation is grappling with a severe shortage of affordable housing, and the need for solutions has never been more urgent. Factors like rising material costs, supply-chain disruptions, and labor shortages have compounded the issue, resulting in a growing deficit in housing availability. While the number of US families increased by 1.8 million between 2021 and 2022, the housing deficit in the US grew to an alarming 4.5 million homes[60]. For many Americans, particularly the younger generations, homeownership has become increasingly unattainable due to this imbalance, which has directly contributed to the worsening affordability dilemma.

Policymakers from both political parties have taken notice of the widening housing shortfall. Despite their many ideological disagreements, the White House and the Republican National Committee (RNC) agree on one thing: selling surplus federal land might be a key factor in addressing the lack of affordable housing.

The Emergence of Bipartisan Support

The acknowledgment of the housing crisis by both leading political parties indicates an escalating urgency. The cross-party agreement on selling federal land at a reduced cost for housing construction indicates that legislators are exploring alternatives to conventional methods to tackle the deficit. However, the proposal does have its controversies. The difficulty is in

[60] Hoover, M., & Lucy, I. (2025, May 17). The State of Housing in America. U.S. Chamber of Commerce. https://www.uschamber.com/economy/the-state-of-housing-in-america

harmonizing conflicting ideas regarding how this program might be executed. Environmentalists are particularly cautious about the chance that this could turn into a handout for developers, worsening environmental issues, and overlooking local communities.

Local zoning regulations, which differ significantly between states, also add to the complexity. Washington has minimal authority over these regulations, and frequently, cities and counties encounter considerable challenges in approving housing projects. This dissatisfaction with local government authority has led to demands for decisive measures, and the concept of using federal land is viewed as a method to overcome several of these obstacles.

A New Vision for Housing: Freedom Cities

Although supply and demand issues have dominated conversations on both sides of the political spectrum, a crucial topic has been overlooked: the necessity of updating home construction methods. In this sense, President Trump's idea of "Entitling Freedom Cities on Federal Lands" has the potential to revolutionize the situation. In addition to selling land, the concept of "entitling" aims to expedite the regulatory procedures that have historically slowed down house developments. These new cities might be built more quickly by avoiding laborious rules and approval procedures.

This vision demands using other measures than the outdated sight-building approach, which frequently leads to waste, delays, and cost overruns. Alternatively, factory-built and modular homes could become the main focus. These homes, constructed in controlled factory settings and then assembled on-site, would provide an affordable and efficient solution to the housing crisis. To draw an analogy, imagine buying a car where the parts are shipped to your driveway and assembled by a crew—an inefficient and cumbersome process. Henry Ford revolutionized the automobile industry with the assembly line, a similar transformation could occur in the housing sector.

From a national perspective, we don't need to eliminate the existing housing market. Rather, we simply need to allow for a new sector within the existing market capable of funding our new Freedom Cities.

The Geography of Freedom Cities: Building in the Arid West

Most federal land where these proposed Freedom Cities could be built is located in the arid regions of the Western United States, notably areas prone to drought. The federal government manages about 28%[61] of the lands within the United States. This is extremely close to 1 million square miles of wilderness. Of this total, the Bureau of Land Management (BLM) manages about 381,875 square miles. Within this vast expanse are areas that could be ideal for developing new urban centers, provided that the design and construction respect the land's unique environmental conditions. The vast majority of opportunities to build new urban centers exist on the BLM lands. The Forest Service manages most of the upper elevations where precipitation falls and trees grow. The Bureau of Land Management has a vast inventory of dryer mountain valleys covered in sage and dry environment grasses. These "fill" valleys host aquifers, streams, and rivers. These mountain valleys host those federal lands that contain the best locations for developing new population centers.

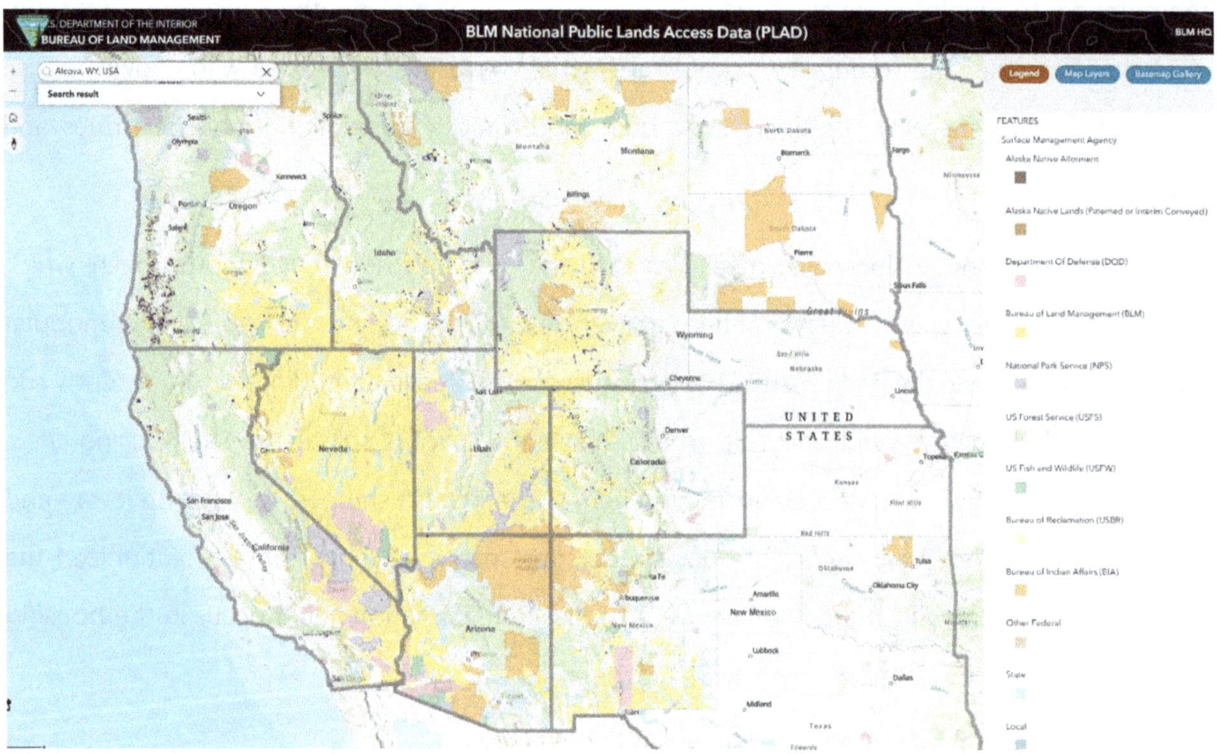

[61] Laura, A. H., Vincent, & Hardy, C. (2020, February 21). Federal Land Ownership: Overview and Data. Congress.gov. https://www.congress.gov/crs-product/R42346

Building in these regions necessitates a delicate balance—keeping the footprint of each city compact while maximizing vertical growth. This approach would not only minimize the environmental impact but also take advantage of modern technology, such as steel-framed modular skyscrapers that could rise up to 32 stories. These modular units, resembling large "LEGO" blocks, could be transported via highways and assembled efficiently, allowing for rapid urban development even in the most remote areas.

Determining the desires and needs of the local populations is a step not yet taken by the proponents of affordable housing on federal lands. Using other companies that build wood-framed modular buildings on a ground floor concrete foundation similar to a parking garage or even your local Walmart would allow for four or five floors of housing above street front shopping and in-structure parking. Building below-the-ground Data Centers with much smaller footprints than the 1-5 square mile cities would achieve economies of scale through cost sharing and provide a boost to the local economy and local jobs. An important concept irrevocably tied to Freedom Cities is to provide economic conditions required to restore the bottom rungs of the latter to success for our youth, veterans, struggling families, and any US citizen willing to relocate to the "frontier" and work to build a common and more prosperous future for all of us.

The inclusion of high-density housing attracts funding, businesses, and a workforce. The degree of local success will depend on the existing communities, the degree to which they are willing to accept some form of manufacturing, and how well they can provide local synergies with existing economic drivers like farming, ranching, mining, and petroleum production. Data centers and power production facilities like Natural Gas power plants and Small Modular Reactors (SMRs) should be considered in the mix of economic drivers.

The Critical Importance of Water

John Westley Powell was a Civil War veteran and the second head of the United States Geological Survey (USGS). He was a famous explorer of the Western United States. In 1879, he published his report to Congress entitled "*Land of The Arid Region of the United States*." In this report, he repeatedly and insistently proclaimed that the water in this region should not be separated from the land. After his death, the Colorado River Compact and other legislation were ratified by the

US Congress, which established the concept of Pueblo Water Rights. The effect of Pueblo Water Rights is that the vast majority of precipitation and run-off that occurs within the higher elevations of the arid region now belongs to the lower elevation coastal cities. It is important to understand that the ratification of such legislation occurred at a time when a military invasion of the US West Coast was considered to be a "clear and present danger." Many of the long and expensive water conveyance systems present to this day in California were financed and constructed to achieve sufficient local populations to protect the US West Coast from invasion. This was a major portion of the political justification for passing such spending measures. The threat of such a military invasion of the US West Coast no longer exists. China lacks sufficient nuclear-powered naval vessels to project power beyond Taiwan, let alone Hawaii or the US West Coast. The supply chain required to provide diesel fuel to the screening vessels would be beyond their near future capabilities. Such an attempted invasion would also be grounds for nuclear war.

One of the most significant challenges when building in the arid regions of the US is the availability of water. These areas often suffer from water scarcity, which could pose a serious obstacle to the sustainability of any new city. The existence of outdated infrastructure, laws, and legal opinions that reinforce antiquated concepts does not help. For Freedom Cities to thrive in these regions, they must prioritize water conservation and management strategies. It will be crucial to design systems that minimize water conveyance length and outdoor usage while ensuring that the needs of residents and businesses are met. This must include the existing needs of our highly valued Farmers and Ranchers. Water will be a critical resource, and without careful planning, the promise of these new cities could falter.

Building High-Density Cities: A Sustainable Approach

One of the most promising solutions for this land is the creation of high-density cities with small footprints, which would respect the land, water, natural resources, and the people who live on the land. By building vertically and focusing on efficient design, it is possible to create cities where affordable housing can be made available to a larger portion of the population. These cities would not necessarily feature sprawling suburban homes with white picket fences, but they would offer individuals the opportunity to own a home and generate equity through property ownership.

This approach could allow for the construction of thousands of units in relatively compact spaces. For example, a building footprint of 4,000 square feet on a piece of land measuring 27.3 square miles could support around 4,182 units, depending on the design and density. Assuming an average of 2.5 residents per unit, the population of such a city would be more than 1 million individuals. This model provides a pragmatic solution to the housing shortage while being in line with principles of sustainability and efficiency.

Selection Criteria for Freedom City Sites

Finding the best places for Freedom Cities requires consideration of strategic, economic, and environmental factors. This will help ensure that the land utilized as a starting point resource could fuel innovation and sustainable growth.

1. Federal Lands as the Foundation for Freedom Cities

A key reason for establishing Freedom Cities on federal properties is the substantial decrease in expenses related to acquiring land. Federal lands, particularly in the Arid Region of the United States, are frequently not fully utilized and can be developed at a much lower cost than acquiring private property. Using federal lands offers a distinct chance to lower costs associated with land acquisition (often a significant expense in construction, particularly in large cities), thereby contributing to more affordable housing.

Moreover, the entitlement procedure, which entails securing approval for development, can be expedited on federal lands, since these regions are frequently already designated for particular uses. A significant rise in population in these areas might also alter the balance of power in the US House of Representatives, possibly affecting national political dynamics. Currently, liberal policies promote higher population density in cities, frequently causing overcrowding that contributes to poverty, crime, and greater reliance on government assistance programs. Developing on federal lands could reverse this trend by distributing populations into less populated areas, leading to more equitable growth.

2. Proximity to Resources and Utilities

To optimize the success and viability of Freedom Cities, sites that are proximate to key resources and facilities are most important. Water, for example, is a particular issue in the dry areas of the United States. Locating Freedom Cities adjacent to available sources of water saves on costly water conveyance infrastructure. Rather than transporting water to the city, the city is constructed near the water.

Aside from water, roads and rail are fundamental parts of any successful city. The Arid Region already has a large interstate system, and the forest service roads exist in many areas. These roads can be paved to increase connectivity between Freedom Cities and existing cities. In the same way, the US rail system is still an effective means of transporting commodities like steel from the local production plants to housing production plants. This streamlined model of transportation, similar to that of Henry Ford's assembly line, would save on construction expenses and accelerate home production.

In addition, construction close to high-tension electric transmission lines is beneficial. This proximity lowers the cost of electricity distribution, ensuring that Freedom Cities have access to reliable power. The natural gas pipeline infrastructure within the Arid Region is also robust, providing an additional energy source for heating and industrial needs.

Building new power plants near new freedom cities will help strengthen the grid and provide good-paying jobs for the local residents. Both natural gas turbine plants and small modular reactors should be considered for strengthening the grid and providing good-paying jobs.

Lastly, the region's airports—especially those within the Colorado Plateau and the Colorado River Basin—remain underserved by regular commercial air services. Building commercial airports capable of servicing 737-sized aircraft would connect these areas to the broader economy, allowing for business, tourism, and trade.

Republican Commitment to Housing Affordability

The vision for Freedom Cities aligns with broader goals to reduce costs and increase opportunities for Americans. Republicans have laid out a plan to make the American Dream more affordable, particularly in terms of housing. Key elements of this plan include:

- Reducing mortgage rates through policies that slash inflation and create incentives for first-time buyers.
- Opening portions of federal lands for new home construction, thus expanding housing availability.
- Promoting homeownership through tax incentives and reducing regulations that unnecessarily increase housing costs.

Additionally, there are commitments to make higher education more accessible and affordable and to tackle the growing problem of healthcare costs by increasing transparency and competition. This multi-faceted approach reflects a commitment to improving the quality of life for all Americans, particularly those who are struggling to afford basic living expenses.

Investment Grade Site Selection Process

The vast majority of federal lands in the US exist within the Arid Region. An investment grade housing project in a high desert mountain valley with a river running through it would reduce the total cost per "condo" built on federal lands within the Arid Region. Properly reclaiming the water used for domestic purposes (indoor water use) would dramatically reduce the downstream impacts on water availability and quality. Irrigation and other forms of (outdoor water use) would need to be limited in these locations. For the average homeowner in the West, the ratio between outdoor water use compared to indoor water use is about 65% to 35%. More than 90% of outdoor water use evaporates or evapotranspiration away in the form of water vapor. More than 90% of indoor water use can be reclaimed by advanced water reclamation systems into class A or A+ effluent water. As discussed in the previous chapter, NASA has recently announced that it is reclaiming 98%[62] of all water consumed in the international space station. Most people fail to understand that "scrubbing the water molecules" to the relatively high-quality class A or A+ effluent water is the easy part. On Earth, streams, riparian ecosystems, aquifers, and even lakes complete the final portion of the process for us. Raising the water quality of discharge from Freedom Cities to something close to drinking water standards and letting Mother Nature take control of the final step is generally considered an environmental benefit and is a far more economically viable solution.

While water availability is the primary risk or limiting factor in producing large-scale "affordable housing" in the Arid Region, site selection and the risks associated with such developments can be managed. This means that federal taxpayers do not need to foot the bill for affordable housing projects. The creation of a "federal backstop" fund that would function to ensure that no such project would run out of money prior to the completion of any given project may be useful. However, the time of NGOs and other organizations receiving grants or any form of public funding to create affordable housing should end or be sharply curtailed. The Freedom City concept detailed here illustrates that we are capable of building affordable housing without "*Big Brother*" getting in the way.

Building a large number of multi-story condominium complexes on top of ground floor shopping, light commercial use, or simply a parking garage has been in practice for generations.

---

[62] ISS: Astronauts have recycled 98% of water from their pee, sweat and breath. (2023, June 27). BBC. https://www.bbc.co.uk/newsround/66017072

Henry Ford invented the assembly line. Today's consumer accepts nothing less than a factory-built car. Lower costs and higher quality are generally associated with any product produced by a factory. Why does the public still lean towards site-built homes?

Water Availability and Legal Challenges

While the potential for large-scale residential and light commercial development is clear, the most significant challenge remains water availability. Though rivers flow from nearby mountain ranges and carve channels in valley floors, the valley floors where much of this land lies do not receive sufficient precipitation to support traditional agricultural or grazing activities. Water rights laws further complicate the situation, as they tend to favor states, cities, and even farms further downstream, such as Nebraska and those on the West Coast, often at the expense of those near the headwaters like Wyoming, Idaho, Southeastern California, Western Nevada, Western Utah, and others.

A noteworthy case regarding water rights occurred between Wyoming and Nebraska over the North Platte River. For years, the two states were engaged in legal battles over the distribution of water from the river, which flows from Wyoming into Nebraska. In 2001, after a prolonged and expensive lawsuit, both states reached a settlement. The agreement included limits on irrigation in Wyoming, increased water monitoring, and allowed Nebraska to retain its right to 75%[63] of the river's water flows. Although the settlement ended the conflict, it also brought attention to the region's complicated water rights laws and the difficulties faced by any major development that depends on a steady supply of water.

Reclaiming Water for Sustainable Development

To address water shortages in regions such as Wyoming, Idaho, and other Western states, the creation of Small Modular Freedom Cities (SMFCs) must incorporate creative solutions such as Advanced Wastewater Reclamation Systems (AWWRS). These systems would enable efficient water reuse within urban areas, significantly lowering the reliance on external water supplies. The concept is that water can be recovered at a rate of 90% for alternative applications like irrigation, landscaping, and certain industrial processes.

The advantages of a one-square-mile city built on this concept are significant. By utilizing reclaimed water for all irrigation in landscaping, these cities would greatly alleviate pressure on natural water resources. With nature's own form of water purification through percolation into the aquifer, this approach also reduces evaporation, a key concern in the arid regions of the US.

These advanced reclamation systems are already being implemented in places like Los Angeles, where the Hyperion Advanced Water Purification Facility and the Donald C. Tillman Water

[63] U.S. Water News Online. (2001, April 20). Nebraska Preserves Water Flows Under Agreement with Wyoming. Wastewater Digest. https://www.wwdmag.com/home/news/10916547/nebraska-preserves-water-flows-under-agreement-with-wyoming

Reclamation Plant provide recycled water for non-potable uses. These systems can even make their way into Freedom Cities, ensuring a reliable and sustainable water supply, even in regions where natural resources are scarce.

Moreover, the integration of modular housing systems, such as Fullstack Modular, Z Modular, or Guerdon Modular, combined with advanced wastewater reclamation systems, will make it possible to create self-sustaining communities in some of the most water-scarce regions of the US.

Desalination: A Possible Solution

For Freedom Cities to thrive in the arid regions of Southeastern California and Western Nevada, another possibility could be the use of desalination technology, which could be employed to convert seawater into potable water for the existing coastal cities in California. This process could help mitigate the impact of limited freshwater resources and provide the water needed to sustain both residents and industries. Desalination, however, is a costly and energy-intensive process. The process of moving water from California's Bay Delta is also costly and energy-intensive. For the coastal cities of California, from San Francisco to San Diego, the cost associated with "going diesel" has never been more affordable. Secondly, the existing water conveyance systems are utilized at a high rate relative to their total capacity in places like the underground pipelines under the City of Los Angeles. **In the recent fires, the hydrants failed due to an overreliance on imported water**. Advocates for desalination and regional self-sufficiency have been silenced by the deep state in California for far too long.

Additionally, the reclamation of water then recycling and reusing it within the city infrastructure could be another way to conserve precious resources while ensuring that Freedom Cities remain sustainable over the long term. Water efficiency, along with careful management of both groundwater and surface water resources, would be essential for the success of these new urban centers.

Pueblo Water Rights in California

Water rights are central to the successful development of Freedom Cities, particularly in the arid regions where these cities are most likely to be established. In California, the concept of Pueblo rights plays a key role in understanding water access and usage. These rights were granted to cities that were founded as Spanish or Mexican pueblos, allowing them to claim all naturally occurring surface and subsurface water from the entire watershed of the stream that flows through the original pueblo.

Pueblo rights are paramount, meaning they take precedence over other water rights like riparian and appropriative rights. Cities that have pueblo rights, such as Los Angeles and San Diego, have a legal right to water prior to statehood and therefore have an upper hand when it comes to obtaining water for their expanding populations. For instance, Los Angeles's explosive growth during the 19th and 20th centuries was in large part driven by the exercise of its pueblo water rights, including rights to obtain water from the Los Angeles River and groundwater in the San Fernando Valley. This special historical entitlement enabled the city to acquire water decades before other water sources, like the Owens Valley Aqueduct, were built in 1913.

However, for new Freedom Cities being developed in areas like Wyoming or Idaho, where pueblo rights do not apply, other water management strategies, including water reclamation and efficient use of limited resources, will be key to sustainability.

Modular Housing and the Affordable Housing Crisis

As the debate over water rights continues, another critical aspect of Freedom Cities' development lies in the construction method. A prominent success in contemporary construction is the creation of 6th Street Place, a modular housing initiative that provides affordable living in Los Angeles. This initiative showcases the ability of modular building to swiftly and effectively tackle the housing shortage, especially for at-risk communities.

Finished in July 2023, 6th Street Place is a six-story structure featuring 93 residential units, aimed at offering permanent supportive housing for those facing homelessness in the Skid Row region. The initiative received partial funding of $14.58 million from the Housing and Community Development Department, contributing to a total budget of $54 million[64]. Using modular

[64] Sharp, S. (2019, November 18). First Glimpse of $54-Million Supportive Housing Project in DTLA. Urbanize LA. https://la.urbanize.city/post/first-glimpse-54-million-supportive-housing-project-dtla

construction, the units were manufactured off-site at Guerdon Modular Buildings in Boise, Idaho, and subsequently put together on-site. This approach not only shortened the total construction duration but also enabled improved quality, providing more cost-effective and durable homes compared to traditional building methods.

Scalable housing options that can be swiftly expanded are possible with modular housing. These modular homes are perfect for Freedom Cities, where effective use of resources and land is crucial, but they can also be positioned in places that do not have the immediate resources for traditional residential building structures.

The NIMBY Mentality: A Barrier to New Housing Development

Despite the increasing demand for affordable housing, a continuous challenge confronting housing development throughout the United States is NIMBYism (Not In My Backyard). The environmental movement of the 1970s helped reduce pollution and improve public health. It opposed harmful projects like the toxic waste site at Love Canal and fought LA's dirty air. Today, that same energy is often used to fight new housing. In California, the Environmental Quality Act (CEQA) requires a full review of a project's impact on the environment before it can be approved. Although this has resulted in some favorable effects, it has also exacerbated the housing crisis by introducing bureaucratic obstacles that hinder or obstruct the building of new residences.

This resistance to change has merely intensified the housing crisis. In cities such as Los Angeles, where soaring living expenses are pushing numerous residents into a challenging predicament where salaries do not match inflation, the creation of new housing has become more challenging to accomplish. As housing prices keep soaring, the demand for additional homes, particularly budget-friendly ones, has become increasingly critical.

The core problem is straightforward: the national housing market lacks 4.5 to 7 million homes[65], and the existing system is not designed to tackle this deficiency. The NIMBY attitude usually stems from anxiety about change, including worries about heightened traffic, pressure on community

---

[65] Hoover, M., & Lucy, I. (2025, May 17). The State of Housing in America. U.S. Chamber of Commerce. https://www.uschamber.com/economy/the-state-of-housing-in-america

resources, or fears of declining property values. Nonetheless, the absence of new housing poses a significantly greater long-term issue for the country overall.

Freedom Cities – A Better Future of America

The Freedom Cities program is more than a reaction to the housing crisis. It is a visionary solution to urban development in some of the nation's most underdeveloped areas. By targeting federal lands in the Arid Region, the idea leverages enormous, unpopulated areas that are not being used to their full potential for housing or community expansion. By leveraging the latest technologies, such as modular construction, water recycling systems, and proximity to existing energy infrastructure, there is a chance to create affordable and sustainable cities. These cities can ease the pressure on overpopulated urban centers, providing economic and cultural renewal to rural and underdeveloped areas.

Chapter 7

Building Multiple Freedom Cities May Depend on Fixing California

The US is also experiencing an acute shortage in housing, with a projected 4.5 million homes[66] needed to satisfy the demand. Thankfully, areas such as Wyoming, Owens Valley, Idaho, and several others have enough space to accommodate about 2 million housing units "without breaking a sweat." This presents an unparalleled opportunity to address this shortage through the development of self-sustaining, purpose-built communities: Freedom Cities. These cities, which are designed to foster job creation, promote individual liberty, and provide affordable housing, could be the solution to America's urban development challenges.

Much of the water needed to build new Freedom Cities is allocated for export to California Coastal Cities. The Colorado River Aqueduct alone moves about 1.2 million acre-feet of water per year[67] from environments and localities that could use the water to achieve a far higher level of development than currently present.

Building Freedom Cities in the Arid Region will require the use of advanced water reclamation plants. Fixing California will require large-scale desalination. Using the existing system of aqueducts to integrate these younger technologies with the older water conveyance system

---

[66] Ibid.

[67] Colorado River Aqueduct. (n.d.). American Society of Civil Engineers. https://www.asce.org/about-civil-engineering/history-and-heritage/historic-landmarks/colorado-river-aqueduct

appears to be the commonsense solution to the housing crisis and long-term agricultural issues facing the region. We should stop talking about removing dams. We should seek to reduce the number of new dams required to fix California's broken water system. The newer technologies will be less expensive and more reliable.

Owens Valley: A Strategic Blueprint for Freedom City

The Owens Valley was transformed in the early 20th century through controversial land acquisitions by the City of Los Angeles. Through a combination of force and manipulation, Los Angeles secured the water rights necessary to feed its growth, diverting hundreds of thousands of acre-feet annually through the Los Angeles Aqueduct. This aqueduct currently delivers approximately 300,000 acre-feet[68] of freshwater to the city each year, with a capacity that can reach up to 480,000 acre-feet in wetter years. The water diversion has hurt the people and environment of Owens Valley the most. But for a long time, leaders have downplayed these impacts. Instead, they've focused more on growing cities than on treating all regions fairly.

A similar story unfolds in Northern California, where San Francisco draws about 265,000 acre-feet annually[69] through the Hetch Hetchy Aqueduct system, supplying water to 2.6 million people across 29 Bay Area cities. Meanwhile, the fertile Central Valley and rural agricultural zones continue to bear the brunt of urban water diversion, leaving many smaller communities water-stressed and economically stagnant.

The recent wildfire season, especially after the January 13, 2025 windstorm, has made clear that Los Angeles can no longer depend on its sprawling water systems to deliver adequate protection. It is in this context that Owens Valley emerges as a prime candidate for the development of a large-scale, master-planned city—Trump's Freedom City. Unlike the overstressed, under-defended urban basins of the coast, Owens Valley offers ample space, historical infrastructure,

---

[68] Owens Valley Water History (Chronology). (2008, January). County of Inyo California. https://www.inyowater.org/documents/reports/owens-valley-water-history-chronology/

[69] Hetch Hetchy Water and Power System. (2024, August 14). Maven's Notebook. https://mavensnotebook.com/explainers/where-does-californias-water-come-from/hetch-hetchy-aqueduct/

and symbolic power. Building here would not only redirect growth away from climate-vulnerable cities like Los Angeles and San Francisco but also create an opportunity to repair past injustices. By investing in modern infrastructure, including large-scale desalination at the coast and inland water reclamation, California could return a significant share of its freshwater supply.

For environmentalists who oppose such ambitious development, it is worth considering the sustained environmental degradation already inflicted on Owens Valley. If California is to transition to a sustainable and equitable future, the answer is not to preserve failed systems but to reimagine them. Owens Valley, with its geographic positioning, existing aqueduct infrastructure, and symbolic history, offers a blank slate upon which that vision can be built.

By locating Freedom City in Owens Valley, California would not only relieve pressure on its crumbling urban cores but also acknowledge and begin to redress the environmental and economic exploitation that has shaped its history.

Wyoming: The Ideal Location for Freedom Cities

Wyoming, the second least densely populated state in the United States, offers vast stretches of land with only six people per square mile. This low population density—combined with the state's abundant natural resources makes it an ideal location for the development of new urban communities. Wyoming's current population of approximately 587,618 residents[70] is relatively small, and much of the land is either mountainous or protected by national parks and monuments, limiting the space available for development.

However, the lower-lying mountain valleys of Wyoming offer significant potential for urban growth, providing ample space for the establishment of Freedom Cities on federal lands.

Strategic Location: Lower Lying Mountain Valleys

The most promising areas for the development of Freedom Cities in Wyoming are its lower-lying mountain valleys, particularly those situated along the "Wind" and "Bighorn" rivers. These river

[70] U.S. Census Bureau QuickFacts: Wyoming. (n.d.). Census. https://www.census.gov/quickfacts/fact/table/WY/PST045224

valleys are well-suited for development due to their reliable water sources. Although they receive less than 14 inches of precipitation annually, the surrounding mountain ranges provide large rivers and aquifers that can meet the water needs of growing urban areas.

One special benefit of these valleys is that they receive less precipitation than higher mountain regions, which is important for urban development because excessive precipitation can make infrastructure planning more difficult. These valleys' rivers, like the North Platte River, supply essential water supplies that allow cities to expand without depleting already-existing resources. Importantly, water disputes are already happening downstream in Kansas, especially between urban and rural consumers. This emphasizes how crucial it is to use cautious water management techniques in any upcoming urban expansion.

Sustainable Water Management

The establishment of Freedom Cities in the river valleys of Wyoming offers a more sustainable approach to water management compared to trying to create large urban centers in the state's dry areas. The river valleys of Wyoming receive under 14 inches of liquid precipitation annually, categorizing the region as semi-arid, but it nonetheless has dependable water sources from rivers and aquifers. These aquifers and surface water can sustain prolonged urban development, guaranteeing that the cities do not encounter the same water scarcity issues as those located in arid areas. The key challenge is to recover most of the water and reintegrate it into the natural system without overwhelming it.

Instead of supplying water to urban areas, constructing Freedom Cities in locations near existing natural water sources is much more logical from both economic and environmental perspectives. This approach lowers infrastructure expenses and alleviates ongoing water management issues. By positioning cities in regions rich in aquifers and surface water, Wyoming can guarantee that these cities access sustainable resources without endangering the depletion of those resources or harming the environment.

Economic Advantages: Jobs, Energy, and Industry

Establishing Freedom Cities in Wyoming's river valleys will also generate substantial economic prospects. These locations, which presently lack excessive urban infrastructure, have the potential to become vibrant economic centers. The energy needs of these cities can be satisfied by utilizing Wyoming's plentiful natural resources, such as geothermal energy and natural gas, which may be employed to offer heating and cooling for the cities in a very energy-efficient manner.

In addition, the state's current energy infrastructure could accommodate emerging industries, like data centers and manufacturing, that need significant electrical power. With the ongoing expansion of the tech industry, data centers will be essential in backing the national drive for artificial intelligence and other advanced technologies. Given Wyoming's cooler temperatures and access to dependable power sources, the state has the potential to emerge as a prime spot for data centers and technology-related businesses.

The cooler temperatures in Wyoming present a perfect chance to harness natural gas or small modular reactors (SMRs) for electricity production. These energy sources can effectively reutilize the waste heat to heat local homes and small cities. Data centers also supply an ample supply of waste heat that could be used for the same purpose. These methods enhance energy efficiency relative to traditional techniques such as the theoretical **Rankine cycle**, which transforms merely around 29% of waste heat into electrical energy. By directly utilizing excess heat for heating applications, Wyoming can achieve a significantly more effective and sustainable energy system.

These new cities could also provide ample job opportunities in manufacturing, agriculture, and energy production. The combination of Wyoming's natural resources and its relatively low cost of living makes it an attractive destination for businesses seeking to relocate or expand.

The Energy Industry and Land Opportunities

In 2023, Wyoming continued to play a critical role in the nation's energy production. The state ranked 8th nationally in both crude oil and natural gas production, making significant contributions to the US economy through property and severance taxes. With its extensive energy reserves, Wyoming remains a key player in the oil and gas sector. By 2022, the state reached a peak of 27,951 producing wells[71], which included both oil and gas wells. Additionally, Wyoming operated 33 gas plants that processed nearly 97% of the state's natural gas production.

Of Wyoming's 23 counties, 21 produce oil and/or natural gas, underlining the state's widespread involvement in energy extraction. Converse County stands out as the leader in crude oil production, while Sublette County is at the forefront of natural gas production. This robust energy sector not only fuels the nation's power needs but also generates substantial tax revenue that supports state and local economies.

Wyoming's vast tracts of land, particularly those managed by the Bureau of Land Management (BLM) and state agencies, could serve as ideal locations for the development of Freedom Cities. The BLM-managed lands in the region are extensive, and even a relatively small portion of land alone could help address the US housing shortage of 4.5 million units[72]. Wyoming is rich in natural resources, which would support the infrastructure needed for new cities.

Adjacent to BLM land, there are areas of state-owned land, further expanding the space available for development. The possibilities for growth in these regions are immense, yet the major hurdle remains water availability. There's a lot of land ready for use, but making these areas into successful cities takes careful planning. A key aspect of this process is ensuring there is enough water for everyone and making sure the supply will meet future needs.

Advanced Water Reclamation: Sustainable Urban Development

The other most important component of constructing Freedom Cities in Wyoming is combining cutting-edge water reclamation technologies. Just like the wastewater treatment plants in California, including the Hyperion Advanced Water Purification Facility, Wyoming's new cities can adopt the latest wastewater recycling systems. These recycling systems would recycle wastewater for purposes other than potable use, such as irrigation and industrial applications, to decrease the dependency of the cities on freshwater supplies.

---

[71] Oil and Gas Facts & Figures 2024. (n.d.). Petroleum Association of Wyoming. https://pawyo.org/facts-figures

[72] Hoover, M., & Lucy, I. (2025, May 17). The State of Housing in America. U.S. Chamber of Commerce. https://www.uschamber.com/economy/the-state-of-housing-in-america

The Hyperion Advanced Water Purification Plant in LA has the goal to recycle 100% of its wastewater[73]. Recycled water in freedom cities can promote sustainable water supply management and reduce dependence on imported water. The most important use of water supplied to these cities is returning most of the water to the natural system.

These advanced reclamation systems could also play a crucial role in supporting agriculture, a key part of Wyoming's economy. The reuse of water for agricultural purposes would allow the state to maintain its food production capabilities while also supporting urban growth. As these technologies become more widespread, they will help ensure that Wyoming's Freedom Cities can thrive in a way that conserves natural resources and sustains local ecosystems.

Navigating Land Ownership and Tribal Issues

While Wyoming offers significant potential for the development of Freedom Cities, challenges related to land ownership and jurisdiction must be addressed. Tribal lands in Wyoming are "held in trust" by the federal government for the benefit of Native American tribes, which can lead to disputes over land use and governance. The Wind River Indian Reservation, which spans parts of central Wyoming, has seen legal disputes regarding the reservation's boundaries and the rights of non-tribal residents in areas like Riverton. In all such conflicts, we must acknowledge that each American citizen deserves access to freshwater.

In 2013, the US Environmental Protection Agency (EPA) determined that Riverton is included in the Wind River Indian Reservation for air quality assessments. This ruling, issued under the Clean Air Act, granted the Northern Arapaho and Eastern Shoshone tribes "Treated as a State" (TAS) status, allowing them regulatory authority over environmental issues in the area. Nonetheless, it also generated apprehensions among non-tribal residents who worried about possible effects on property rights, taxes, law enforcement, and governance.

In reply, the state of Wyoming contested the EPA's ruling in the 10th Circuit Court of Appeals, with the case staying unresolved for several years. In the meantime, Northern Arapaho Business

---

73 Hyperion Advanced Water Purification Facility (CA). (n.d.). Water Collaborative Delivery Association. https://watercollaborativedelivery.org/project/hyperion-advanced-water-purification-facility/

Council Chairman Darrell O'Neal highlighted the significance of open communication, encouraging stakeholders to attend a symposium called "*Mending Fences: How One Michigan Town Resolved Its Boundary Battle Without the Courts.*" The event sought to encourage conversations about collaborative solutions instead of lengthy legal disputes.

The dispute highlighted a wider trend in federal policy choices concerning Native American land rights and management. The legal viewpoint of the Obama administration in 2013 represented an inappropriate enlargement of tribal authority, similar to other regulatory overextensions like the "Green New Deal." Such actions were viewed as politically driven attempts to seize power instead of sincere attempts to resolve historical land conflicts. On the other hand, proponents of the decision saw it as a long-needed acknowledgment of tribal sovereignty and legal rights as defined by federal law.

President Trump and his team, in collaboration with the Department of Justice (DOJ), are focused on minimizing government waste, fraud, and abuse. Tackling concerns such as the Wind River conflict would necessitate thoughtful negotiation and tactical measures ahead of the 2026 midterm elections. As political momentum shifts from left to right, many see this as a rare opportunity to create a more balanced long-term approach to governance. However, deep-seated government bureaucracy, especially in places like California, has often proved resistant to federal intervention.

Trump has suggested that all American citizens, regardless of race, color, creed, sex, or religion, have the right to own and use condos or other forms of housing built in freedom cities. Furthermore, ensuring that plenty of jobs are made available to sustain the American dream for future generations.

Although these disputes are complex, they should not deter the development of Freedom Cities in the region. With proper planning and respect for tribal sovereignty, these issues can be resolved in a way that benefits all parties involved and allows for the development of sustainable, thriving communities.

Political Hypocrisy and Bureaucratic Overreach in California

The image of California Governor Gavin Newsom dining maskless at the exclusive French Laundry restaurant during the height of the COVID-19 pandemic became an emblematic moment of political hypocrisy. At the time, California residents were under some of the strictest public health orders in the nation, including mask mandates, curfews, business closures, and limits on gatherings, imposed by the very officials who were not following them. Newsom's behavior, photographed, also generated widespread public outrage, not only for the obvious contradiction but because it highlighted a greater problem: increasing disconnection between California's ruling elite and its ordinary citizens.

Californians have been complaining more and more about what they perceive to be a two-tiered society in which elites and politicians follow a distinct set of laws. Residents started to doubt how to strike a balance between governmental control and public safety as a result of rigorous regulations that many thought violated their personal freedoms and severely restricted small enterprises and places of worship. The French Laundry episode wasn't isolated; rather, it came to represent a larger trend in governance that was characterized by a lack of consistency, overzealous regulation, and a perceived decline in public confidence.

Economic Manipulation

California's economy is strong enough to endure and shape national trends. Nonetheless, this impact has frequently been utilized to sway the economy toward state objectives:

- Enforcing rules that limit farmers' water usage directly results in supply limitations and increased food costs.
- Financial support and tax breaks are often allocated to projects that coincide with the prevailing political agenda, such as robust green energy initiatives.

The issue commonly called the "green new scam" is rooted in this overwhelming bureaucratic and political oversight, directing public funds and resources into selectively advantageous programs while neglecting wider economic well-being.

The Principle of "Rules for Thee, But Not for Me"

The idea of elitism and double standards is perhaps the most obvious problem highlighted by events such as Governor Newsom's French Laundry dinner. Political elites frequently break the law with impunity, despite the fact that regular people are supposed to abide by stringent laws.

This hypocrisy fosters public resentment and undermines trust in government institutions. The concept of "rules for thee, but not for me" encapsulates a culture where accountability is selective and power is exercised without genuine responsibility.

To what extent should the President and DOGE control California's substantial federal funding, and what portions could potentially be eliminated? The election of a Republican governor like Steve Hilton would certainly help, but he would need to win in a landslide and gain multiple seats in the state's house and senate in order to get anything of substance done.

If the President proposes building SMRs (Small Modular Reactors) and desalination plants managed by federal contractors operating fully in line with DOGE protocols, the estimated cost to "fix California" — at least in the water and power sectors — could cost around $161 billion. This amount closely matches the federal funds allocated to California, which in the fiscal year 2022 totaled approximately $161.7 billion. These funds supported critical infrastructure, education, healthcare, and public welfare. We all have some idea of how much waste is associated with this funding in the state of California. California's share made up 12.9%[74] of all federal transfers to states that year. California's existing political and cultural environment has been influenced by years of progressive policies and social beliefs, many of which originate from the early 1990s. Certain regulatory foundations date back to 1850, the year California became the 31st state in the Union.

Through the years, these policies have significantly shaped the state's trajectory, integrating ideas of Diversity, Equity, and Inclusion (DEI) with contentious environmental positions.

---

[74] How Much Federal Money Goes Toward California State and Local Government? (n.d.). USA Facts. https://usafacts.org/answers/how-much-money-does-the-federal-government-provide-state-and-local-governments/state/california/

The rise of radical leftist beliefs and DEI-focused practices started to gather pace in California in the late 1980s. The presence of a real communist party in the country is extensively recorded in the years following WWII. Since that time, these frameworks have greatly influenced the political environment as well as cultural and educational institutions throughout the state.

A notable example is the hosting of events like Drag Queen Story Hour, designed to captivate children with concepts of reading and gender expression. While proponents see these actions as acknowledgments of diversity and inclusion, opponents contend they signify a more profound cultural change driven by "woke left" ideas that are not well-supported by facts.

An essential aspect of stabilizing California involves reassessing how youth are educated. Instead of instructing children on what to believe, equipping them with the skills to think critically would foster a more resilient and reflective democratic republic. By having a more robust grounding in critical thinking, upcoming generations would be more prepared to tackle societal issues and oppose ideologies that could lead to negative outcomes.

Fixing California: Politics First, Technology Follows

The first and most important step in solving California's problems is political reform. No technological advancement—whether in desalination, energy, or infrastructure will suffice without tackling the deeply rooted political structures perpetuating these issues.

Fortunately, solutions like desalination technology are readily available "off the shelf," awaiting implementation. But real progress begins with fixing the politics, setting the stage for broader, more sustainable improvements across the state.

Understanding How The Existing Infrastructure Works

Looking at the extent and average annual flows within the major aqueducts that feed California's coastal cities is an important step in understanding how to fix many of the issues that have plagued the state for decades.

The illustration and accompanying legend should provide a basic overview of California's major aqueducts.

Integrating Desalination into The Existing System

This may be the most important issue covered in this book. If we choose to desalinate large amounts of freshwater from saltwater in close proximity to California's coastal cities and feed this water directly to the cities only, the cost of creating this new infrastructure will be minimized. The need for new dams and reservoirs will be all but eliminated. More of the water stored in the existing dams can be redirected to other uses like farms and a new Freedom City or two.

(Socal Water Conveyance Infrastructure)

For San Francisco, the three main aqueducts (in acre-feet/year)

- 195,000 Contra Costa aqueduct[75]
- 364,000 Mokelumne aqueduct[76]
- 265,000 Hetch Hetchy aqueduct[77]

824,000 Combined delivery to San Francisco

For Los Angeles through San Diego

- 1,100,000 The California aqueduct
- 300,000 The Los Angeles aqueduct
- 1,200,000 The Colorado River aqueduct

2,600,000 Combined delivery to LA through San Diego

Agricultural deliveries to the Southern Central Valley

- 4,556,000 Water removed from the Delta by the CA Aqueduct
- -1,100,000 Less Water delivered to LA-San Diego
- 1,051,000 Water from the Friant Kern Canal[78]

4,605,000 Agricultural deliveries to the Southern Central Valley

Water delivered to the Imperial Water district and local municipalities near the Salton Sea

- 3,100,000 Water from the All-American Canal[79]

Approximate deliveries to cities: 3,554.000
Approximate deliveries to Agriculture 7,705,000
California does not have sufficient water for both.

---

[75] Contra Costa Water District Water Services. (2008, April 9). Contra Costa LAFCo. https://www.contracostalafco.org/municipal_service_reviews/central_county_water_wastewater/4.0%20CCWD%20Water%20Wastewater%20Final.pdf

[76] Mokelumne Aqueduct. (n.d.-b). Maven's Notebook. https://mavensnotebook.com/explainers/where-does-californias-water-come-from/mokelumne-aqueduct/

[77] Hetch Hetchy. (n.d.). In Wikipedia. Retrieved February 16, 2025, from https://en.wikipedia.org/wiki/Hetch_Hetchy

[78] Friant-Kern Canal. (n.d.). In Wikipedia. Retrieved November 3, 2024, from https://en.wikipedia.org/wiki/Friant-Kern_Canal

[79] All-American Canal | Imperial Irrigation District. (n.d.). Imperial Irrigation District. https://www.iid.com/water/water-transportation-system/colorado-river-facilities/all-american-canal

The Desalination Debate – Environmental Myths vs. Urgent Realities

The Propaganda Against Coastal Desalination

Critics, particularly those aligned with environmental alarmism, frequently argue that coastal desalination poses a threat to local marine ecosystems. However, this narrative is largely propaganda. Their true interest lies in maintaining the status quo, the so-called "gravy train" of entrenched policies and outdated systems. Global data consistently demonstrates that modern coastal desalination technologies are environmentally safe and are widely used by leading nations without the catastrophic impacts often claimed.

Urgent Priorities for California

Rather than focusing on speculative environmental concerns, California should prioritize addressing the state's very real and ongoing environmental and economic crises. Modern desalination plants present a viable, immediate solution to California's water scarcity without significant ecological harm.

The current failure to act reflects a broader issue: The Government of California has neglected its fundamental responsibility to safeguard its citizens' lives, liberty, and property. The comparison data showing how the UAE and Saudi Arabia have effectively scaled desalination efforts should serve as a wake-up call and a point of embarrassment for policymakers in California. The UAE provides 2.25 million acre-feet per year of desalinated water to its citizens, and Saudi Arabia provides 2.35 million acre-feet per year. California provides less than 100,000 acre-feet per year to its citizens.

Replacing about 2-million-acre feet per year of freshwater imported to the coastal cities of California with desalinated water along the coast would solve two large issues. First, it would provide a redundant source of water for the coastal cities and reduce the possibility of fire hydrant failures in the future by creating a redundant supply. Second, it would create larger supply to all inland water uses including building new Freedom Cities and agricultural uses.

California's Water Revenue, Spending Concerns, and the Road to Political Change

Water Revenue and Household Costs

California has approximately 13,217,586 households, with an average household size of 2.92 people. On average, residents pay about $65 per month for water services. With roughly 90%[80] of households relying on public water systems, the state generates substantial revenue from water bills:

- Monthly Water Revenue: ~$773 million
- Annual Water Revenue: Over $9.2 billion

This revenue supports a water infrastructure system that, for the most part, was paid off decades ago (excluding maintenance costs). Despite this, the state continues to collect billions each year from its residents.

Where Does the Money Go?

The allocation of this massive revenue stream raises critical questions:

- How much of the $9 billion collected annually is lost to waste, fraud, and abuse?
- How much is diverted to NGOs or political patronage networks?
- In 2022 alone, California spent $7.2 billion on initiatives for the homeless population — how much of that was truly effective?

It's unlikely that California's persistent water issues will be solved without the expansion of coastal desalination. Privately operated desalination plants could help rebalance the water supply while reducing the financial strain and political entanglements surrounding state-controlled water services.

[80] Water Systems. (2025). Tracking California. https://trackingcalifornia.org/topics/water-systems

Desalination as a Solution for Bay-Delta Restoration

The reliance on the Mokelumne, Hetch Hetchy, and Contra Costa aqueducts significantly reduces freshwater flow through the Bay Delta. These outdated systems divert water away from critical ecosystems, impacting the health of the Bay-Delta, Suisun Bay, and connected rivers.

By replacing a substantial portion of these freshwater flows with desalinated water, California could pursue much-needed environmental restoration for these waterways. This approach would restore natural flows, support local habitats, and reduce ecological damage caused by excessive diversions.

Proposed Desalination Initiatives

Port Chicago Naval Base — Low Salinity Desalination Plant

- Target capacity: ~400,000 acre-feet of desalinated water annually
- Focused on providing desalination of low-salinity water to offset inland freshwater demands.

Pillar Point Air Force Base — Ocean Water Desalination Plant

- Target capacity: ~200,000 acre-feet of desalinated water annually
- Designed to utilize ocean water and reduce dependency on freshwater aqueducts.

Vision for Restoration

Implementing these desalination projects would allow California to:

- Reduce the strain on the Bay Delta and its tributaries.
- Revitalize ecological conditions in the Bay Delta and Suisun Bay.
- Modernize the state's water infrastructure while ensuring long-term sustainability.

Innovative Diffusion Line Strategy for Low-Salinity Desalination

The strategy is to collect the brine concentrate, the byproduct of reverse osmosis, and release it strategically during outgoing tides to minimize environmental impact.

How the System Works

- The lower section of the diffusion pipeline is perforated (represented by the red arrows in the image).
- This perforation allows the brine to diffuse as it is discharged during outgoing tides, using natural tidal currents, especially the high-turbulence areas in Carquinez Strait, to aid dispersion.
- Tidal currents here naturally create turbulent eddies, enhancing the mixing of the brine with surrounding waters and reducing localized salinity spikes.

Understanding the Concept — An Analogy

To visualize this:

- Imagine an Olympic-sized swimming pool.
- If you add fresh water to one end while adding a gallon of saltwater at the other, the impact on overall salinity is minor because of the pool's size and water movement.
- Similarly, in this system, outgoing tides act like the water flow in the pool, helping dilute and disperse the saline concentrate downstream, especially in areas like San Pablo Bay.
- Measuring changes in salinity downstream helps monitor and ensure environmental safety.

While this is a simplified analogy, it effectively captures the concept of strategic timing and dispersing brine discharge to minimize ecological disruption.

Strategic Water Management for the Bay Delta and Port Chicago Desalination

Port Chicago Desalination Plant – A Salinity Control Solution

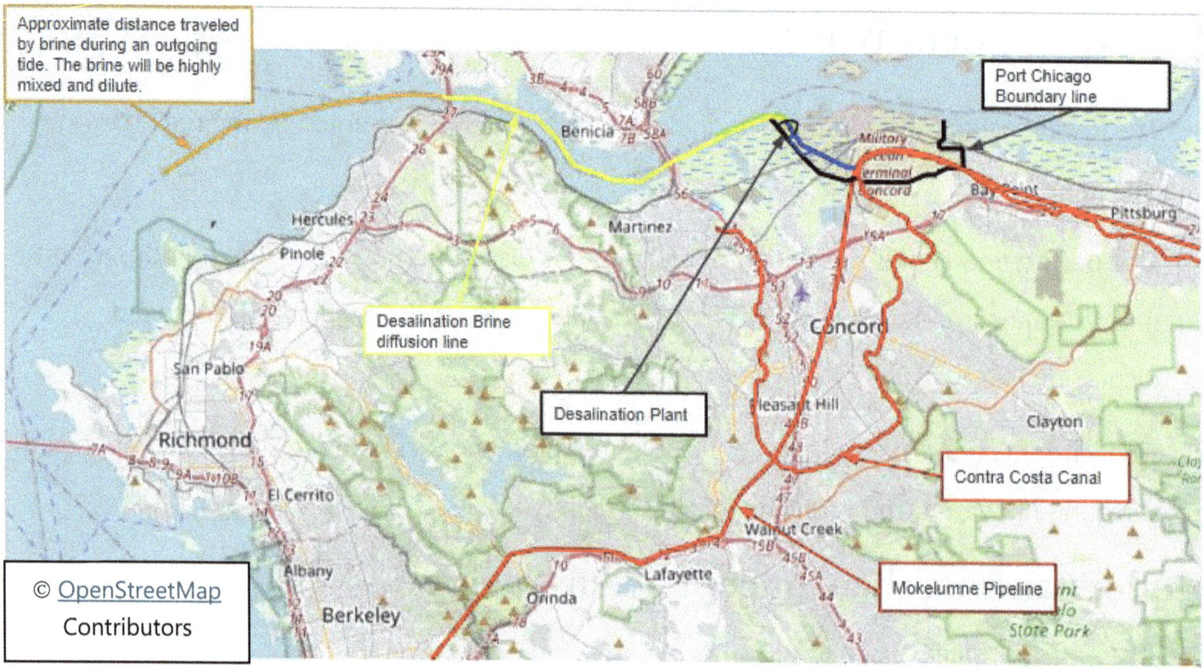

(The Port Chicago Desalination Plant)

The Port Chicago desalination plant is designed to serve as a freshwater source and a protective barrier against saline water intrusion into Suisun Bay. By reducing the amount of freshwater directed out to sea, we can better maintain the freshwater balance in this sensitive region and prevent higher salinity levels from encroaching into the Bay Delta ecosystem.

Enhancing Freshwater Flow to the Delta

Additionally, approximately 600,000 acre-feet of freshwater annually will continue flowing into the Delta, sourced from:

- Eastern regions
- Dams at the headwaters of the Mokelumne and Hetch Hetchy aqueducts.

These water releases must be timed strategically, ideally aligning with late spring when delta flows are typically at their lowest. This timing ensures maximum efficiency in sustaining delta water levels and ecological health.

The "New Twin Tunnels" Project

A key infrastructure development is the "New Twin Tunnels," which will:

- Desalinating low-salinity water at Port Chicago is measured at approximately 5 parts per thousand (ppt), compared to ocean water at 35 ppt. This significantly reduces energy consumption and operational costs.
- The lower salinity results in less saline brine rejection from reverse osmosis (RO) membranes, minimizing environmental impact and allowing for a more flexible and efficient system design.
- The Mokelumne and Contra Costa systems currently divert a combined total of approximately 1,018,000 acre-feet of water per year. Reintroducing this volume into the natural system and instead desalinating low-salinity water offers a more sustainable approach.
- New desalination infrastructure would support increased water exports during wet years, enhance salinity control in the Delta Suisun Bay, and contribute to the protection of sensitive species such as the Delta Smelt.
- The two 8-foot diameter pipelines would enable the transfer of approximately 1 million acre-feet of water annually to Clifton Court Forebay.
- The two 10-foot diameter pipelines would increase this capacity to 1.6 million acre-feet annually.

- The strategic location of new intake structures allows for a more diversified and adaptive water intake system that preserves natural flow patterns within the Delta during both wet and dry seasons.
- This infrastructure model offers improved regulatory control over freshwater discharges to the ocean versus deliveries through the California Aqueduct.

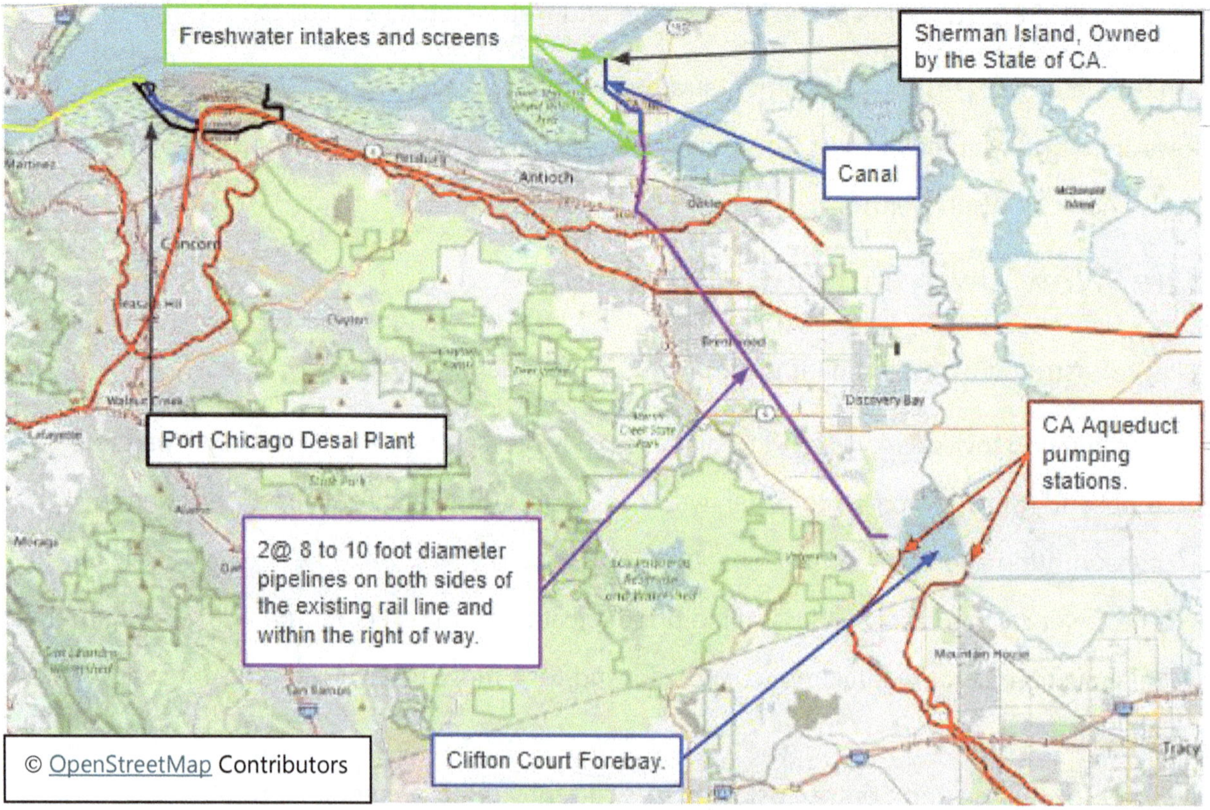

(The New Twin Tunnels)

Revising the Bay Delta Conservation Plan (BDCP) for Sustainable Water Management

Original BDCP Proposal

The initial Bay Delta Conservation Plan (BDCP) put forth the building of two colossal tunnels, one 40-foot wide and spanning 30 miles long[81]. They were intended to channel water from the Sacramento River through environmentally delicate locations with a mission to preserve the Delta system and provide for a reliable water supply for California's economy.

Improved Design: A Smarter Alternative

The revised design offers a more effective and eco-friendly option:

- The new alignment is 16.5 miles long, significantly shorter than the original proposal.
- It is optimized to make use of "un-diverted" freshwater from the Mokelumne and Hetch Hetchy aqueducts, improving the natural water movement into and across the Delta.
- The design incorporates 9-foot-diameter steel pipes instead of large tunnels, making it less disruptive and better aligned with the ecosystem.
- The system features an enhanced saline water rejection mechanism that contributes to healthier salinity levels in the Delta and nearby ecosystems.

Utilizing Excess Water from the Mokelumne and Hetch Hetchy Aqueducts

Building on findings from the referenced research, the water not exported through the Mokelumne and Hetch Hetchy aqueducts can be strategically used to boost California's agricultural water exports during the critical late spring through early fall irrigation period. This approach would sustain "exports from the Bay Delta" at higher levels throughout the growing

[81] Delta Conveyance Project. (n.d.). In Wikipedia. Retrieved April 14, 2025, from https://en.wikipedia.org/wiki/Delta_Conveyance_Project

season. About 5% of water exported from the Delta is used for irrigation[82]. To meet irrigation demands, at least 1.5 million acre-feet of water should be exported from the Delta annually.

However, the existing system tends to "crash" during peak irrigation periods. By providing an additional:

- 1.2 million acre-feet of water per year for the Southern Central Valley farmers, and
- 300,000 acre-feet annually for urban use in Los Angeles.

California can unlock multiple benefits.

Key Benefits of This System:

- Lower overall food costs
- Boost the economic and environmental health of the Delta and Central Valley.
- Increase monsoonal rainfall in the arid Southwestern U.S.
- Support greater population growth in arid regions
- Fuel the growth of freedom cities (decentralized urban development)
- Transform national supply and demand dynamics in U.S. housing
- Reduce long-term inflationary pressures
- Revive the American Dream, making homeownership and prosperity more attainable.

Important Note:

Average ocean salinity is approximately 35 Parts Per Thousand (PPT)[83].
For effective ocean water desalination, water must be pressurized ("pumped up") to force it through the reverse osmosis (RO) membrane, which separates salt molecules from water molecules.

---

[82] The Sacramento– San Joaquin Delta. (2016). In Public Policy Institute of California. https://www.ppic.org/wp-content/uploads/content/pubs/report/R_1016JM5R.pdf
[83] Ocean Water Properties. (n.d.). Hurricanes: Science and Society. https://www.hurricanescience.org/science/basic/water/index.html

The resulting "reject brine" from the desalination process is:

- Typically, twice as salty as the intake water, around 70 PPT, and
- Warmer due to the pressure and energy applied.

By contrast, water at the green intake line proposed for the Port Chicago plant is estimated to have much lower salinity, around 5 PPT.

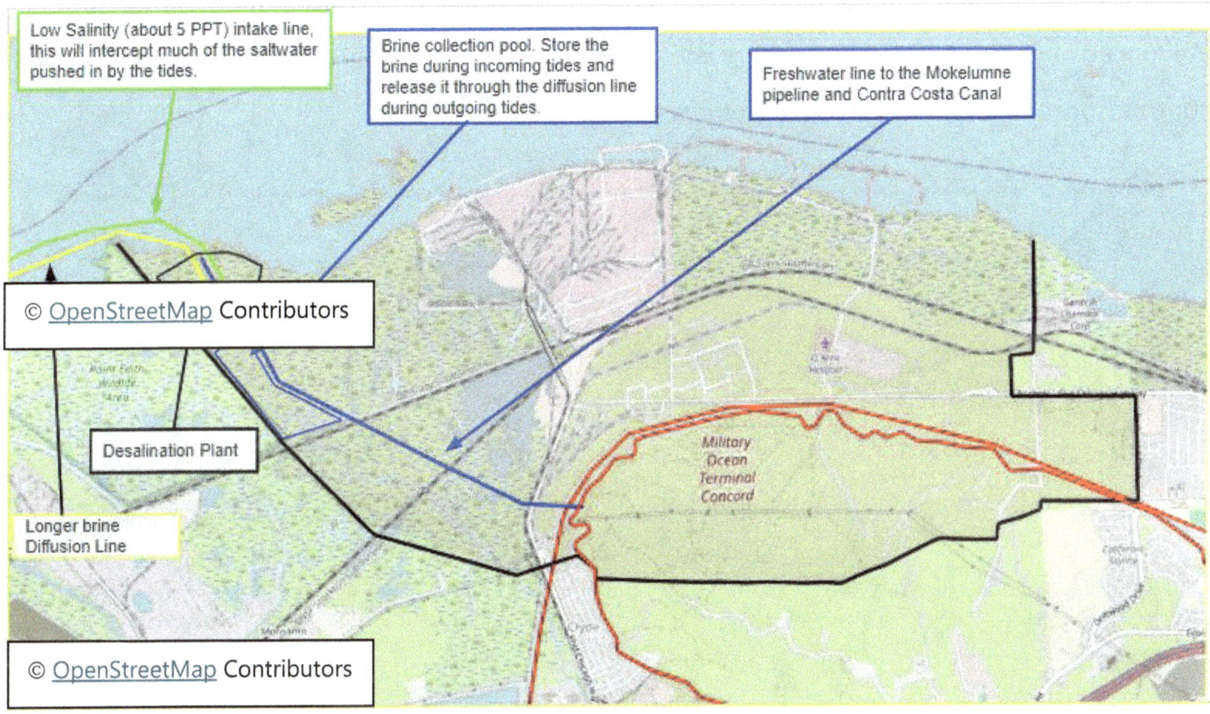

(The Port Chicago Desalination Plant)

This lower salinity provides clear advantages:

- Reduced operating costs, as less energy is needed to process the water.
- Higher efficiency, producing more freshwater per unit of input water.
- Lower salinity in the brine, depending on the system pressure applied.

However, there are trade-offs:

- Higher pressure systems allow more water to be desalinated, but also result in warmer, saltier brine output.

- Engineering solutions are needed to design and place plants like the Port Chicago desalination facility to minimize environmental impact.

Fortunately, the technology for desalination and advanced water treatment is rapidly maturing, making these challenges increasingly manageable.

It's also important to remember that the U.S., and particularly California, is not the first region to depend on desalination for water security. However, many policies in California—especially those influenced by bureaucratic inefficiencies ("CCCP")—have been irrational and are a major contributor to failures like *"Why the Hydrants Failed."*

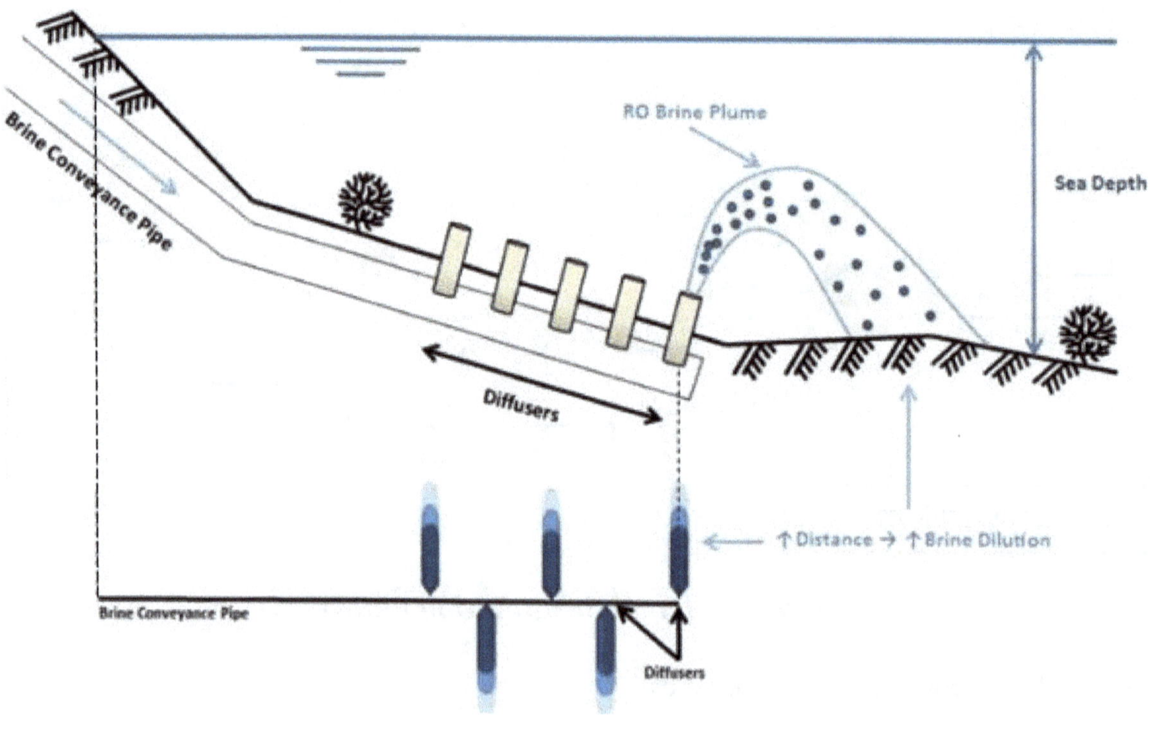

(The Port Chicago – Brine Conveyance Pipe)

Expanding Desalination Capacity in Southern California

In Southern California, spanning from Los Angeles to San Diego, there is significant potential to establish at least five major desalination plants. These facilities could be built with minimal environmental impact, especially if they are strategically placed on military bases, where open land is still available. Given the scarcity of undeveloped land in the region, military installations represent some of the last viable locations for such critical infrastructure.

As mentioned earlier (see Slide 20), a target of at least 600,000 acre-feet per year of locally produced desalinated water is recommended to support the region's water demands.

Specifically, for the LA to San Diego corridor, I recommend focusing desalination efforts to:

- Maximize local water independence.
- Reduce reliance on imported water sources.
- Strengthen water security for urban centers and surrounding communities.

The 2026 Gubernatorial Election: A Chance for Change

California's next gubernatorial election, scheduled for November 3, 2026, with the primary set for June 2, represents a critical inflection point for the state. With Governor Gavin Newsom term-limited and unable to seek re-election, voters have a rare opportunity to bring fresh leadership into Sacramento. More than a dozen candidates have already entered or expressed interest in the race, many promising to tackle the state's chronic challenges from failing infrastructure and housing shortages to the ever-worsening water crisis.

The 2026 race could mark a turning point where Californians demand more than slogans. Voters are growing weary of expensive mandates with no quantifiable results, performance politics, and bloated bureaucracy. The need for policies based on common sense, openness, and long-term sustainability as opposed to ideological extremes or virtue-signaling is increasing.

Political Influence Over Climate Narratives

For many years, California has led the way in influencing national policy directions, especially in terms of environmental regulation and climate change. Nonetheless, a significant portion of this impact has originated from established political and bureaucratic organizations—some detractors compare them to CCCP-style think tanks, unselected panels, and international NGOs that craft narratives without substantial public responsibility. These organizations frequently promote policies labeled as "climate action," yet the actual impacts on average Californians, such as elevated energy costs, closed businesses, and more regulatory hurdles, are often overlooked.

The so-called "green new scam" appears to focus less on safeguarding the environment and more on strengthening political and financial control. These ideological frameworks seldom accept opposing viewpoints, even when supported by scientific facts or historical background.

Misguided Incentives in Climate Science

Science must never be used as a political instrument. Nevertheless, in the current research climate, numerous scientists are under economic pressure to deliver findings that fit politically popular agendas. Governmental and private sources of funding tend to favor studies confirming fears of increasing CO2 and global warming, dissuading open-ended or dissenting research.

To true scientists, this subverts the spirit of discovery and the seeking of truth. When research becomes a race for grants, not answers, the credibility of the entire scientific community suffers. It also marginalizes important findings that challenge prevailing views but could enrich the public's understanding of Earth's natural systems.

Contextualizing CO2 and Global Temperatures in Earth's History

The Earth's climate history spans roughly 4.5 billion years. In that context, today's atmospheric CO2 levels—hovering around 420 parts per million are still low compared to many prehistoric eras. During parts of the Mesozoic and Paleozoic periods, CO2 levels were several times higher, and life on Earth flourished. Likewise, current global temperatures are modest when compared to ancient warm periods such as the Eocene Thermal Maximum.

Climate activists often cite that *"95% of scientists agree"* that CO2 contributes to global warming. While technically accurate, this statistic oversimplifies the issue. Yes, CO2 is a greenhouse gas, but its role must be understood within the larger, dynamic framework of solar cycles, ocean currents, volcanic activity, and natural feedback loops. Present-day warming trends may, in fact, reflect Earth's slow recovery from an extended glacial epoch rather than a crisis.

Far from being a pollutant, carbon dioxide is essential to life. Over the past 35 years, satellite imagery has documented a dramatic global greening effect, driven largely by rising CO2 concentrations. According to NASA and NOAA, 25–50% of vegetated land worldwide has seen increased leaf area, particularly in arid and semi-arid regions. This "CO_2 fertilization" has enhanced plant growth, improved crop yields, and expanded the habitat for wildlife.

In truth, warmer temperatures and higher CO_2 levels often lead to more abundant life—not less. This is not to dismiss environmental stewardship but rather to call for a more balanced, data-driven conversation about climate and energy policy—one that values human progress alongside conservation.

Any politician claiming that CO_2 emissions will trigger the collapse of human civilization within the next 10,000 years deserves skepticism and perhaps should be dismissed from office for spreading such unfounded fears. Earth's climate system, often described as "spaceship Earth," has sustained life for approximately 4.5 billion years, weathering far more severe climate events than human-induced CO_2 emissions. Consider these natural phenomena:

- Volcanic eruptions and tectonic shifts
- The 11-year solar cycle
- The 26-year lunar cycle
- Asteroid, meteor, and comet impacts
- Solar flares
- Local supernovae and other cosmic events

Each of these has had significant impacts on Earth's climate, influencing factors like ocean evaporation rates, rain distribution, storm intensity, and the positioning of critical atmospheric zones such as the intertropical convergence zone. In comparison, current CO_2 levels are a minor factor within this vast and resilient planetary system.

Chapter 8

Fixing California Is A Political Issue Not A Technical One

California's political landscape has been shaped by a single-party governance system that has actively suppressed the development of nuclear power and desalination along its coast. The decommissioning of the San Onofre Nuclear Generating Station more than a decade ago marked the loss of a significant power generation source. Meanwhile, the state's reliance on traditional hydroelectric power is insufficient to meet the growing demands of its water distribution system. Despite the challenges, California continues to depend heavily on petroleum-based power generation to sustain its existing water infrastructure. At the same time, the state imposes stringent limitations on wind and solar power, leaving residents to cope with the consequences.

A basic message in this book is that we currently have the technology to reclaim and desalinate water in sufficient quantities to continue our economic growth and protect the environment. Perfecting a fusion-based energy source or relying on wind and solar is not required.

The Promise of NuScale Small Modular Reactors

One potential solution that could address both the state's energy and water needs is the use of NuScale small modular reactors (SMRs), which are designed for the co-generation of electricity and desalinated water. These reactors offer an efficient and environmentally friendly way to produce both power and clean water, using nuclear energy to generate thermal energy for desalination without emitting greenhouse gases. Studies show that coupling SMRs with desalination plants, especially those employing reverse osmosis (RO) offers the most favorable economic outcomes. Along with this, the flexibility of SMR design allows for integration with thermal distillation plants or hybrid configurations, providing greater revenue potential and flexibility.

However, the political landscape poses the most significant barrier to deploying SMR-powered desalination plants. The California Coastal Commission, a body dominated by unelected, liberal members, exercises considerable influence over energy projects along the coast.

The Commission's members, appointed by the Governor and other legislative bodies, hold the power to prevent the development of nuclear power and desalination infrastructure. Their political leanings often lead to opposition against projects that involve nuclear power, particularly along the state's coastal areas. The consequences of this political suppression are profound, as they prevent the adoption of solutions that could alleviate California's water crisis and energy challenges.

The Water Monopoly and Its Impact on Innovation

The situation is further complicated by the fact that California's water system operates as a near-monopoly, with government agencies holding significant sway over pricing and policy. This monopoly, in conjunction with the political dynamics, hinders the adoption of innovative solutions, such as nuclear desalination. Ratepayers effectively fund these policies, perpetuating the single-party state's political influence over the water sector. As a result, the very policies designed to address water scarcity may, in fact, be contributing to the state's ongoing water crises.

In this context, President Trump holds unique constitutional powers that could facilitate a solution. As Commander in Chief, he has the authority to take executive action on military bases and could potentially bypass state-level opposition to nuclear desalination plants. Under his oath to protect and defend the Constitution, the President has a duty to ensure the lives, liberty, and property of US citizens, particularly when natural disasters—such as wildfires—threaten communities. As seen in California's frequent fire seasons, a redundant and resilient water system could make all the difference in ensuring that communities have access to water during emergencies. However, California's lack of local water storage and its policy of leaving reservoirs empty as part of conservation efforts undermines the state's preparedness for such disasters.

The Need for a Redundant Water System in Southern California

The state's high-risk areas, particularly in Southern California, require a more effective water management strategy. High property values, combined with the threat of wildfires and earthquakes, make the need for a reliable water system even more urgent. A redundant water infrastructure, which includes properly filled local reservoirs, would provide greater resilience in the face of these natural disasters. Yet, state policies that prioritize conservation over practical emergency preparedness are insufficient.

At the same time, the state's failure to address water issues has led to political disputes. President Donald Trump's threat to withhold federal disaster aid for wildfire-ravaged Los Angeles highlighted the tensions between federal and state officials over water management policies. Trump's comments, blaming the state's fish conservation efforts for hydrants running dry during wildfires, underscore the conflict between water conservation goals and the reality of water scarcity during disasters. The wildfires themselves have had devastating economic consequences, with estimates placing the cost of damages in Southern California between $250 and $275 billion[84].

Integrating State-Owned Water Systems with Private Desalination Plants

Integrating California's state-owned water system with privately owned desalination plants along the coast would present a significant opportunity to restore freshwater flow through the Mokelumne River and the Delta itself. Historically, taxpayers and ratepayers have been the primary

---

[84] Semancik, A. (2025, February 6). The Economics of a Disaster: How the LA wildfires may impact the economy. Ohio University. https://www.ohio.edu/news/2025/02/economics-disaster-how-la-wildfires-may-impact-economy

sources of funding for the construction and maintenance of California's extensive water distribution system. By incorporating desalination plants into the state's water infrastructure, particularly along the coast, California could address water scarcity issues while simultaneously improving freshwater flow in vital river systems.

A critical component of the state's water strategy is the addition of more southern storage. A second San Luis Reservoir site could provide additional storage capacity, enabling more water to be moved from the Delta during the Spring flood season. This increased storage capacity would ensure more efficient water management and provide a buffer during dry periods. Additionally, an optional small tunnel "intertie" system for the Western reservoirs would greatly benefit farmers in the Southern Central Valley by improving water delivery and creating a more integrated water system across the state.

The Need for More Southern Storage Solutions

While some proposals suggest building new or raising existing dams in the northern regions of California, such strategies overlook the opportunities for expanding water storage in the southern part of the state. Politicians who advocate for northern storage fail to recognize the importance of Southern California's water needs. More southern storage options exist, and these should be prioritized as part of an integrated water supply system for California. Southern storage solutions would enhance the state's ability to handle fluctuations in water availability and improve the overall resilience of the water distribution system.

Site Location for Desalination and SMR Plants

The proposed locations for new desalination plants would be capable of producing between 400,000 to 600,000 acre-feet of water per year, coupled with small modular reactors (SMRs) to generate the necessary power. These plants would integrate with existing water sources like the Mokelumne pipeline and the Contra Costa Canal, which together provide significant amounts of freshwater to California residents. The Mokelumne pipeline spans 95 miles and has a 61-inch diameter, delivering water to 1.4 million people[85]. The Contra Costa Canal, on the other hand, is 48 miles long[86] and serves about 200,000 residents with 195,000 acre-feet of water per year.

Existing Water Infrastructure and Its Challenges

The Mokelumne River serves as a crucial water source for the East Bay Municipal Utility District (EBMUD), which has water rights, allowing the withdrawal of up to 364,000 acre-feet of water annually[87]. This water is distributed to 1.4 million residents. Similarly, the Contra Costa Water District (CCWD) receives 195,000 acre-feet per year from the Central Valley Project (CVP)[88], with water quality that can vary significantly depending on environmental conditions. The Contra Costa Canal, which plays a vital role in distributing this water, often faces challenges with water quality, including fluctuations in salinity and organic carbon levels. The integration of desalinated water into these systems could help mitigate water quality issues and improve the overall reliability of the water supply.

The Role of Desalinated Water in Providing Redundancy

Given the high demand for water in these regions, the existing infrastructure operates at near full capacity, leaving little room for maintenance or upgrades. By introducing desalinated water into the system, redundancy could be added to meet the needs of 1.6 million people, including both residential and commercial users. The redundancy offered by desalinated water would allow for better maintenance scheduling and prevent system failures, which are increasingly critical as California's water infrastructure ages and faces growing pressure from climate change and population growth.

---

[85] Mokelumne Aqueduct. (n.d.). Retrieved May 4, 2025, from https://en.wikipedia.org/wiki/Mokelumne_Aqueduct

[86] Contra Costa Canal. (n.d.). Contra Costa Water District. https://www.ccwater.com/212/Contra-Costa-Canal

[87] Mokelumne Aqueduct. (n.d.). Retrieved May 4, 2025, from https://en.wikipedia.org/wiki/Mokelumne_Aqueduct

[88] Contra Costa Water District. (2024). FY25–FY26 Budget. In Contra Costa Water District. https://ccwater.com/Archive/ViewFile/Item/396

The X2 Delta Smelt Protection and Environmental Considerations

In addition to meeting water needs, desalination plants would need to consider environmental protections, such as the X2 targets set by the US Fish and Wildlife Service and the California Department of Fish and Wildlife. These targets are designed to protect the low salinity habitat for the Delta Smelt and maintain it at or less than 80 km from the Golden Gate Bridge in the fall during wet and above-normal water years. As desalination plants are integrated into the water system, care must be taken to ensure that they do not negatively impact these critical ecosystems. The salinity levels in the Delta fluctuate dramatically, from 0 parts per thousand (PPT) during flood events to 12 PPT during droughts and incoming tides. By designing desalination plants to release brine only during outgoing tides, potential negative effects on water quality can be minimized.

The footprint of existing petroleum refineries and power generation plants along the coast provides a valuable comparison of the environmental and economic impacts of desalination and SMR plants. These existing industrial sites offer a practical "side-by-side" analysis for evaluating how the integration of desalination technology could impact both the environment and the economy. The key factors to consider are the environmental impact of intakes, water quality, and the management of brine, alongside the economic benefits of providing a more reliable and sustainable water supply to California's growing population.

X2 and Low Salinity Measurements

X2 is a critical measurement in understanding the salinity levels in the Delta and surrounding areas. X2 targets, specifically the 80 km location[89] from the Golden Gate Bridge, are generally reached during periods of low "out-to-sea" flows from the Delta into Suisun Bay. During wetter periods, larger flows can sustain the X2 distance closer to the 60 km mark. The goal is to collect low-salinity water from the 55-60 km range on the X2 scale, near the main shipping channel at depths of 15-20 feet, where the salinity remains stable between 5 and 10 parts per thousand (PPT).

[89] Salinity. (2025, March 11). Delta Stewardship Council. https://viewperformance.deltacouncil.ca.gov/pm/salinity

This area provides a consistent and reliable water intake source for desalination plants or other water extraction methods.

Ralph D. Bollman Water Treatment Plant

The Ralph D. Bollman Water Treatment Plant, located in central Contra Costa County, is a state-of-the-art facility that provides high-quality drinking water to numerous communities, including Concord, Clayton, Clyde, Pacheco, Port Costa, and parts of Pleasant Hill, Walnut Creek, Martinez, and Bay Point. The plant, built in 1968, has undergone two major upgrades to incorporate the latest advances in water treatment technology. The second upgrade introduced ozone as the primary disinfectant, positioning the Bollman Plant at the forefront of modern water treatment. Ozone disinfection significantly enhances water safety by eliminating harmful microorganisms and breaking down organic materials, improving both the taste and odor of the water. Additionally, ozone treatment helps reduce the formation of potentially harmful disinfection by-products, such as trihalomethanes, which are linked to cancer risks associated with chlorine-based disinfection.

The plant's ability to meet increasingly stringent state and federal water quality standards has ensured that the residents of central Contra Costa County continue to receive some of the safest and cleanest drinking water available. The facility's advanced disinfection process results in water that is free from taste and odor issues while maintaining the highest safety standards for public health.

Tidal Currents and Erosive Flow in Carquinez Strait

The tidal currents within the Carquinez Strait can reach speeds of 5 feet per second or more, a velocity considered to be erosive. This fast-moving water is responsible for maintaining the deep channel in the Strait, which is the deepest part of the entire estuary. In contrast, the shallower depths at the ends of the Strait are more prone to sediment buildup, requiring regular dredging to ensure that the shipping channel maintains its minimum depth for navigation. The federal maintenance project for the Roe Island shipping channel ensures that the channel's depth remains at 30 feet, with a project depth of 30 feet across most of the channel. Reach A of the channel has a controlling depth of 32 feet, while Reach B has a slightly shallower depth of 28 feet.

San Pablo Bay and Suisun Bay

San Pablo Bay is a significantly larger body of water compared to Suisun Bay, with its volume more than double that of the latter. The volume of water exiting this large bay system is approximately equal to the total freshwater inputs from precipitation, runoff, and reclaimed water from treatment plants. Most of the water treated at plants like the Chicago Port Plant is used for indoor residential water use, with a significant portion of that water being reclaimed and returned to the system.

Saltwater Intrusion and its Environmental Impact

Saltwater intrusion into Suisun Bay and the Delta itself is a critical issue that is driven primarily by tides and the movement of cold salty water upstream through the deepened shipping channels. This phenomenon can be compared to the movement of cold air into the US Midwest, where dense, cold air forces thunderstorms and tornadoes. Similarly, saltwater intrusion results from the denser, salty water traveling upstream in the deepened shipping channels, primarily in the Carquinez Strait. The X2 measurement, which tracks the distance of the saltwater intrusion, is typically taken 1 meter above the bottom of the shipping channel to account for the deeper saltwater penetration.

When the flow of freshwater from the Delta to Suisun Bay falls below 4,500 cubic feet per second (CFS)[90], the saltwater intrusion into the Bay increases, pushing the X2 measurement closer to the Golden Gate. The constant monitoring of this data is vital for understanding and mitigating the impacts of saltwater intrusion, which affects both water quality and ecosystem health in the region.

Strategic Use of Water Resources

By strategically managing the intake of low-salinity water from Suisun Bay and exporting the brine further downstream into the Carquinez Strait, it is possible to maintain a consistent freshwater flow in Suisun Bay, thus keeping the X2 point further out to sea. This approach would

---

[90] Assessment of Water Availability Issues Within the Legal Delta. (2021). In Water Unavailability Methodology for the Delta Watershed. https://www.waterboards.ca.gov/drought/drought_tools_methods/docs/wua_app_d_082021.pdf

not only improve the quality of water for treatment purposes but also help mitigate the negative effects of saltwater intrusion on the region's ecosystems and water infrastructure.

Environmental Solutions and Restoring Water Rights

One potential solution to address these issues is the installation of two 10-12 foot diameter pipelines, which could be trenched primarily along existing rail easements. This method would serve to solve many of the environmental concerns currently affecting the Delta, including water rights disputes, and provide a means to restore access to water for local farmers. By utilizing existing rail easements, the environmental footprint of the project could be minimized while allowing for the efficient transport of water.

This plan also has the potential to increase the amount of water exported from the Delta, providing a significant boost to the state's water supply system. Trenching the pipelines along these established easements would help streamline the process, minimizing land disruption and reducing potential opposition from environmental groups and landowners.

California Aqueduct Pumping Stations

In addition to the proposed pipelines, there are multiple state and federal pumping stations strategically located across the California Aqueduct system. These stations serve as the first in a series of pumping stations designed to move water efficiently through the system. By increasing the capacity of the aqueduct and enhancing its pumping infrastructure, California could improve the flow and distribution of water to key regions in the state.

To optimize the system, the concept of trenching pipelines near existing infrastructure, such as rail easements, could facilitate smoother water transport with minimal environmental disruption. The existing 30 to 40-foot rail easements from the outside rail of the tract could serve as an ideal location for laying down the new 10-12-foot diameter pipes on each side, ensuring that water reaches its destination in a timely and environmentally responsible manner.

If properly constructed, these new pipelines will simply add more diversity to the existing system of intakes and restore the natural flow on the deal, all while moving more water to the South.

Addressing Water Access and Exporting Solutions

Tapping into the Sacramento River at strategic points, combined with infrastructural improvements such as the proposed pipelines along existing rail easements, could provide an effective way to solve multiple water-related issues in California. These measures would help restore water rights to family farmers, enhance water exports from the Delta, and ensure a more sustainable water distribution system for the state's growing population and agricultural sector. By integrating both environmental considerations and practical engineering solutions, California can move closer to addressing its long-term water challenges.

Chapter 9

Baja California as a US Protectorate

The Mexican State of Baja California encompasses the entire Baja California Peninsula and has never been part of the United States. While the US gained large swaths of land from Mexico in the mid-19th century, Baja California remained firmly within Mexico's sovereignty. After the Mexican-American War ended in 1848, the Treaty of Guadalupe Hidalgo redrew the borders, ceding what is now California, Arizona, New Mexico, and parts of several other states to the United States. Baja California, however, was retained by Mexico and has remained so ever since.

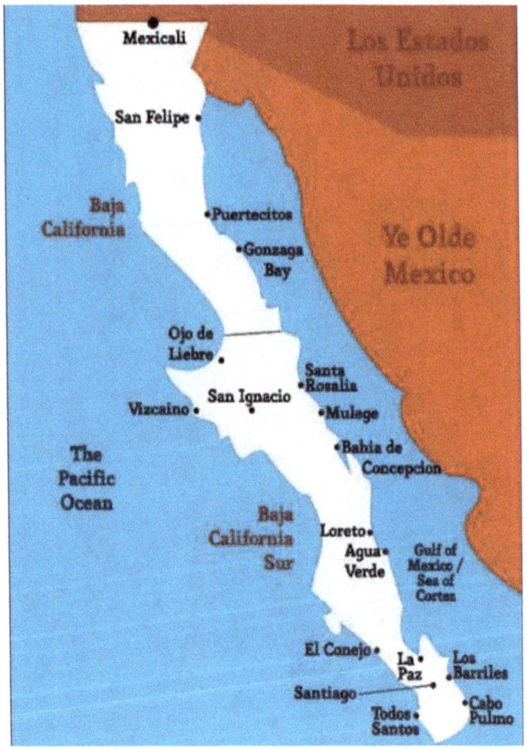

This chapter does not propose a rewriting of that history but rather a strategic reassessment of what the future might look like, particularly if Baja California were to become a US protectorate. Not a state, not an annexation, but a protected territory wherein its people retain their elected government, language, and cultural identity while gaining defense and infrastructure benefits from the United States. Most importantly, such a move could offer protection from the pervasive violence and coercion inflicted by drug cartels, who operate with near impunity across the peninsula today.

Water Scarcity Problems

Water scarcity is one of the most pressing challenges facing Baja California today. Tijuana, the city along the US border, is running out of water options. Desalination has become an urgent necessity. Rigoberto Laborin Valdez, a senior official with Baja California's Public Services Commission (CESPT), was blunt in a press conference shortly after Governor Jaime Bonilla took office. He said that desalination is the only viable solution for Tijuana's long-term survival. Efforts to expand the aqueduct from the Colorado River have been deemed inadequate, and water recycling strategies fall short of the growing demand. Even if water flows from the Colorado River aqueduct increases to between 185 and 190 million cubic meters annually, it will be insufficient in the upcoming years. Tijuana is also responsible for the massive amount of sewage contaminating the infamous Tijuana River.

The limitations of the current emergency water delivery system underscore the crisis. Governed by an agreement signed in 1972 along with several amendments or "minutes" since, this cross-border arrangement has allowed for emergency deliveries of water from US sources to Mexico. Volumes have fluctuated from zero to nearly 10,000 acre-feet per year, though the maximum authorized annual volume is 14,400 acre-feet. These deliveries were consistent through the 1970s, then ceased from 1981 to 2002. They resumed from 2003 to 2012, stopped again from 2013 to 2017, and have occurred almost every year from 2018 to 2024, with the exception of 2021.

The delivery process is administratively complex, involving numerous agencies on both sides of the border. US participants include the Bureau of Reclamation, the US International Boundary and Water Commission, the Metropolitan Water District of Southern California, the San Diego County Water Authority, and the Otay Water District. On the Mexican side, the National Water Commission, Mexico's International Boundary and Water Commission, and the State Public Utility Commission of Tijuana (CESPT), among others, are involved.

The current agreement, set to expire in 2027, outlines procedures for annual requests, capacity assessments by California agencies, delivery cost calculations, and financial guarantees. With each request, Mexico must provide payment in advance, and US agencies must confirm that capacity exists within their systems to honor the delivery. It is a fragile system, both politically and logistically, and it is no longer enough.

The Protectorate Framework as a Viable Solution

A protectorate framework could lead to infrastructure investments currently hindered by political risk. The concept of a protectorate has served as a geopolitical tool to stabilize unstable regions while preserving a degree of local autonomy. Unlike colonization, which demands direct governance and complete administrative control, a protectorate is a partnership—albeit one between an unequal pair. The stronger state offers security, order, and resources in exchange for influence over foreign policy, defense, and economic frameworks. The protected territory retains its cultural identity and often its internal governance, but now within a framework of support that ensures long-term viability. A possible alignment of a new section of border wall from Yuma, New Mexico, to Deep into the Colorado River delta is shown in the illustration below. This would protect "the new protectorate" from hordes of immigrants overwhelming the new Baja California border as they did the US border during the Biden Administration. The wall would be along the thick red line.

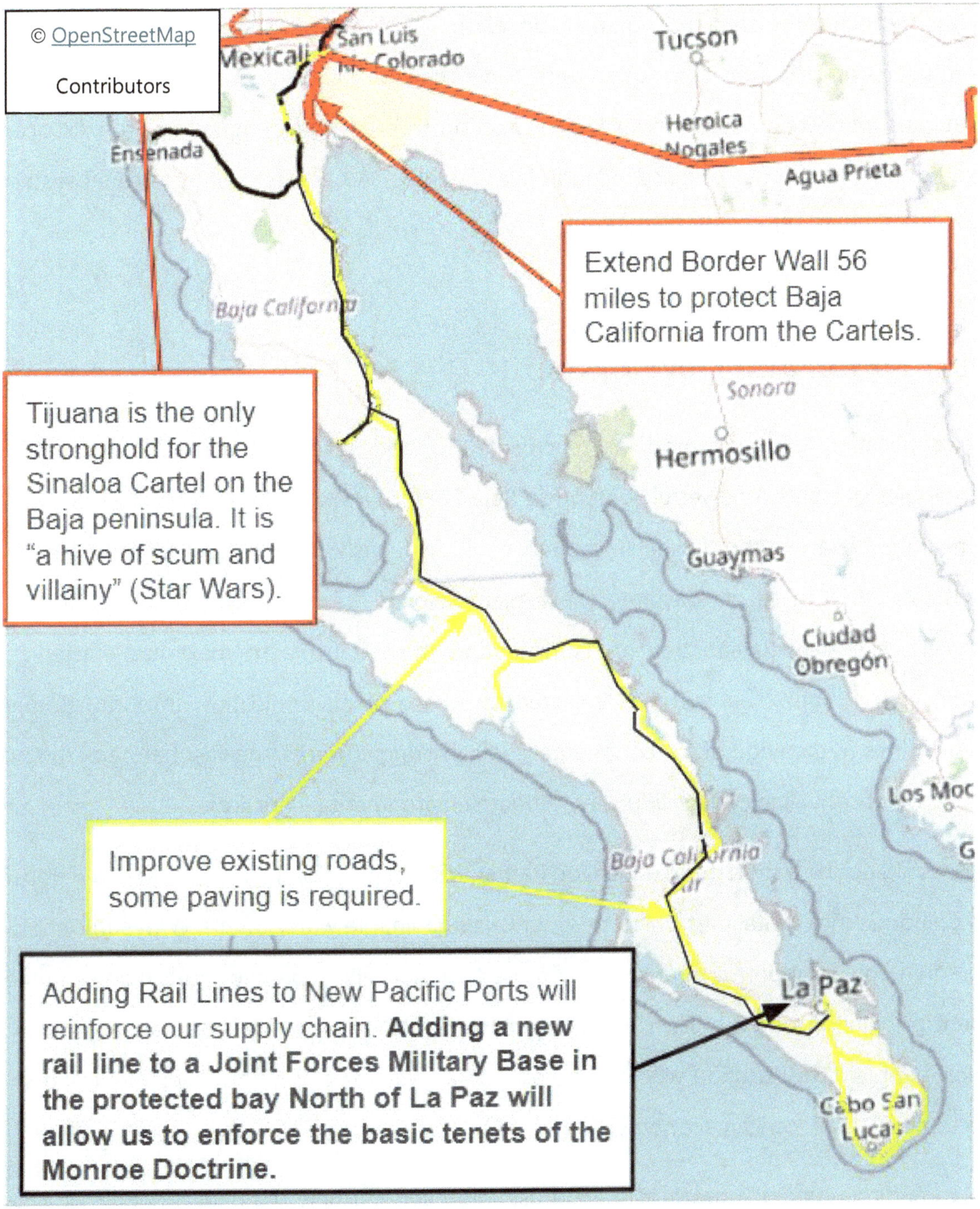

(Baja Roads and Rail)

In the 19th and early 20th centuries, this model was used extensively by European powers in Africa and Asia, often to secure access to resources and markets. While the motives of those imperial powers were far from altruistic, the underlying structure remains a practical and effective model under the right circumstances.

Today, protectorates are rare. But they do exist. The Cook Islands, for example, are a self-governing territory in free association with New Zealand. Palau, a small Pacific island nation, operates as a sovereign state under U.S. protection through the Compact of Free Association. These arrangements offer defense, economic assistance, and access to international systems in return for strategic alignment.

So why not Baja California?

Desalination and Development

The application of small modular reactor (SMR) and reverse osmosis (RO) desalination technologies presents a powerful solution to Baja California's water crisis. The US, leveraging its engineering capacity and financial capital, could establish desalination facilities along the peninsula's 2,038 miles of coastline. These plants would ideally be distributed strategically—especially near deep ocean trenches surrounding the peninsula to maximize efficiency and scalability. The higher salinity brine "rejected" by these plants could flow through submerged diffusion lines to the edge of the continental shelf. Existing currents, the vast nature of the ocean, and the thermo-hyaline cycle will further mitigate environmental concerns.

This would not be a short-term endeavor. Rather, it's a generational project that could transform Baja California into a vital part of North America's sustainable water future. With sufficient water, the region could support a population of more than 40 million people more than ten times its current population of 3.8 million. For perspective, San Diego County alone already hosts 3.27 million people. This location would likely attract large sums of foreign investment and should help to grow the American economy.

Baja California, under a protectorate status, could serve a new national function as a "boiler room" for America's melting pot. Here, immigrants could begin their journey toward US citizenship. They could learn English, study US civics, and be thoroughly vetted before moving into the continental United States. The peninsula could be a proving ground—not just for individuals—but for technology, governance, and border management. Developing Baja would make it into a vacation paradise. An opportunity for US citizens to purchase and own property in

Baja could, at least temporarily, reverse the direction of immigration. Such an event would modestly reduce housing demand in the Western US.

Security and Sovereignty

The most immediate benefit, however, might be security.

Baja California, like much of northern Mexico, suffers under the constant threat of cartel violence. Entire communities live under a shadow of fear, extortion, and displacement. Becoming a US protectorate could offer a transformative shield against these forces. American military presence particularly through a new joint services base at the southern end of the peninsula could disrupt cartel logistics, cut supply lines, and deliver a level of public safety that local forces have struggled to provide.

Moreover, the extension of US coastal waters southward would provide strategic advantages, especially in countering maritime drug trafficking. A naval base across the Gulf of California from Sinaloa (the main hub of the Sinaloa Cartel and a known fentanyl production zone) would place a direct American presence in a region that desperately needs containment and oversight.

For Mexico, this arrangement could be mutually beneficial. Freed from the strain of securing a distant and destabilized territory, Mexico could focus resources elsewhere. Foreign investments that are currently constrained by federal restrictions on property ownership could finally flow into Baja California. Under the protectorate framework, allied nations and corporations could buy, hold, and sell land without the barriers that have stifled development in the past.

The Geopolitical Stakes

In the annals of US foreign policy, few doctrines have shaped hemispheric relationships more profoundly than the Monroe Doctrine. Declared in 1823, its core tenet was clear: the Americas were to be free of European colonial ambitions and interventions. Any foreign interference in the Western Hemisphere would be viewed as a direct challenge against the United States. Over time, this doctrine evolved from a warning to a cornerstone of American grand strategy throughout the 20th century, reinforcing US influence and interests in the region.

Today, a new form of incursion threatens the sovereignty and security of the United States. Chemical weapons disguised as illicit drugs. The rise of synthetic opioids, especially fentanyl, has become an undeclared war on American soil.

The numbers are staggering and sobering. Between 2020 and 2021, overdose deaths involving synthetic opioids, including fentanyl, increased by over 22%[91]. In 2021 alone, nearly 71,000 Americans died from fentanyl overdoses[92]. To put this into historical perspective, that's more than 29 times the number of Americans killed during the Pearl Harbor attack on December 7, 1941, where 2,403 military personnel and civilians lost their lives. During the entirety of World War II, approximately 420,000 Americans died due to war-related causes. And yet, the opioid crisis has claimed more than 1 million lives since 1999.

The scale of this crisis is not only national. It is global in origin but intimately tied to regional dynamics. The chemicals required to manufacture fentanyl are primarily shipped from China to Mexico. Once there, the cartels, specifically the Sinaloa Cartel and others, synthesize the drug in clandestine labs and smuggle it into the United States through border cities like Tijuana.

These cartels have grown exponentially in wealth, firepower, and political influence. Their profits from both drug and human trafficking operations have outpaced those of many multinational corporations. As cartel revenues have grown, the authority of the Mexican government has eroded in inverse proportion. In some regions, cartel power eclipses that of elected officials and law enforcement.

If the Mexican government has, through neglect or complicity, enabled these conditions to persist, then it has implicitly created an undeclared state of war with the United States. Few nations would tolerate the systematic killing of their citizens by a foreign-adjacent criminal enterprise

---

[91] Spencer, M. R., Miniño, A. M., & Warner, M. (2022). Drug Overdose Deaths in the United States, 2001–2021. In NCHS Data Brief (No. 457). Centers for Disease Control and Prevention. https://www.cdc.gov/nchs/data/databriefs/db457.pdf

[92] U.S. Overdose Deaths In 2021 Increased Half as Much as in 2020 – But Are Still Up 15%. (2022, May 11). Centers for Disease Control and Prevention. https://www.cdc.gov/nchs/pressroom/nchs_press_releases/2022/202205.htm

without recourse. Yet the US response remains fragmented, constrained by diplomacy and geography.

One radical but increasingly logical response would be to reconsider the terms of the 1944 Water Treaty between the US and Mexico. Under this agreement, the United States delivers 1.5 million acre-feet of water annually from the Colorado River to Mexico[93]. These flows originate from major infrastructure points like the Glen Canyon Dam (forming Lake Powell) and the Hoover Dam (forming Lake Mead), both of which are currently at critically low levels due to prolonged drought and overuse.

---

[93] Staff, M. (2025, April 29). Mexico reaches agreement to send more water to southern US. Mexico News Daily. https://mexiconewsdaily.com/news/mexico-us-reach-water-deliveries-agreement-texas-farmers-relief/

(Baja Desal Locations)

Revoking or reducing this water supply would be a severe, potentially lethal, diplomatic action. But in the context of more than 1 million overdose deaths—many caused by fentanyl trafficked through Mexico such a measure would be seen by some as proportionate. Water, like oil in the 20th century, is a global tool. It can nourish, or it can devastate. It can be an instrument of peace or a catalyst for war.

Using Baja California As A Strategic Buffer Zone

Amid this tense geopolitical landscape, the proposal to establish Baja California as a US protectorate gains clarity. The region is already geographically and demographically distinct from mainland Mexico. Separated by the Gulf of California, Baja is a long, narrow peninsula with an arid climate, limited freshwater sources, and a population of just 3.8 million[94]. It is far less conducive to large-scale narcotics agriculture, unlike mainland states such as Sinaloa or Michoacán, which possess fertile valleys for growing marijuana and coca.

With the possible exception of Tijuana, the current cartel activity in Baja, though still present, is comparatively minor. This is in part due to geography and, in part, due to the economic unsuitability for traditional drug crops. However, the synthetic nature of fentanyl manufacturing means that arid regions are no longer immune. The base chemicals imported from China are processed using chemical techniques that often involve water as a universal solvent, which remains critical in these operations. Without intervention, Baja could soon become another battleground.

One potential solution would be to extend the existing border wall southward, physically separating Baja California from the rest of mainland Mexico. This would create a clear demarcation and serve as a buffer zone severing cartel supply chains and limiting the spillover of violence and corruption from the mainland. It would also reinforce Baja California's identity as a semi-autonomous zone, distinct in governance, economy, and trajectory.

Transforming Baja California into a US protectorate would result in various strategic advantages. Much like current US protectorates such as Guam, American Samoa, Puerto Rico, and the US Virgin Islands—the region would remain outside the path to statehood but under the protection and governance framework of the United States. This would allow the implementation of law enforcement reforms, anti-corruption measures, and advanced infrastructure planning without fully integrating the territory into the union.

[94] Statista. (2024, August 21). Mexico: population in Baja California 2008-2022. https://www.statista.com/statistics/1038033/mexico-total-population-baja-california/

Most crucially, such a designation would permit:

- Property ownership by non-Mexican citizens, eliminating one of the primary obstacles to foreign investment.
- Massive foreign capital influx, spurred by reduced sovereign risk and US legal protections.
- Deployment of Small Modular Reactors (SMRs) and reverse osmosis (RO) desalination systems, enabling long-term water sustainability.
- Population expansion, allowing the region to eventually support tens of millions through water independence and security.
- A holding platform for immigrants who could begin the path toward citizenship while undergoing civic education, language training, and comprehensive vetting.

A higher quality of life for current residents, including access to education, healthcare, employment, and safety.

Why Water, Immigration, and Infrastructure Demand a New Border Logic

San Diego County, located in one of the most water-strained states in the US, is home to 3.27 million people[95]. Directly across the border, the Mexican state of Baja California hosts 3.8 million residents[96]. Combined, these regions form a binational corridor of over seven million people relying on shared geography, a strained river system, and radically different governing bodies.

Keep in mind that the United States maintains the single largest consumer market in the world. Entire economies are built around access to this market, with countries clamoring to move goods, services, and people across American borders. That makes what's happening just south of California not only unacceptable but strategically dangerous.

---

[95] Epps, R. (2025, January 31). Navigating San Diego's economic future: Insights from the 41st Annual Economic Roundtable. San Diego Workforce Partnership. https://workforce.org/news/navigating-san-diegos-economic-future-insights-from-the-41st-annual-economic-roundtable/
[96] Statista. (2024, August 21). Mexico: population in Baja California 2008-2022. https://www.statista.com/statistics/1038033/mexico-total-population-baja-california/

Despite the global spotlight on US borders, most of the chaos is funneled through one predictable pathway that we often see footage of migrant caravans on. A line drawn along the central body of Mexico might as well be relabeled the Sinaloa Cartel Illegal Immigration Express. From Mexico's southern border to the US border, migrants, drugs, and criminals are trafficked with industrial efficiency. It's no mystery how they arrive—only a disturbing unwillingness by the Mexican government to stop it.

The threat isn't coming from the Baja Peninsula. It's coming from the mainland. The "main body" of Mexico acts as a corridor of corruption, while Baja California remains largely outside of this dominant trafficking network. That geographic distinction is precisely what makes Baja a viable candidate for a fundamentally new role as a US protectorate.

The Sewage Crisis No One Wants to Talk About

As if a lack of freshwater weren't enough, Baja California exports something else to the US every day—untreated sewage and industrial toxins.

The Tijuana River Watershed is a lifeline for people on both sides of the border and the ancestral homeland of the Kumeyaay people. But today, it's a conduit of contamination. Crippled by a century of mismanagement, the watershed channels raw sewage, chemicals, and debris across the border during seasonal rains and infrastructure failures. Beach closures in San Diego. Ecosystem collapse. Illness in children and families. This is the price of inaction.

Kristan Culbert of American Rivers put it bluntly:

> *"Clean water is a universal need and a shared value. But pollution in the Tijuana River is making people sick. No families should have to worry about getting sick from polluted water, and no children should grow up on polluted beaches. It is long past time for action."*[97]

The most egregious part? The pollution flowing into the US from Tijuana represents a daily, invisible breach of national integrity. If a border wall keeps out people, what protects our citizens

[97] American Rivers. (2024, April 16). Tijuana River Named Among America's Most Endangered Rivers® of 2024. https://www.americanrivers.org/media-item/tijuana-river-named-among-americas-most-endangered-rivers-of-2024/

from waterborne pathogens, heavy metals, and fecal bacteria? The condition of that water is so dire it could be considered another form of illegal immigration—one that crosses without inspection spreads unchecked, and inflicts real harm on American citizens.

A Modern Solution to a Borderland Crisis

Baja California is geographically and strategically unique. It is insulated from the mainland cartel highways, its problems are solvable with the right tools, and its people stand to benefit greatly from infrastructure, rule of law, and shared prosperity. But that only happens if we stop viewing this region through the same outdated lens.

A US protectorate model wouldn't be a hostile takeover. It would bring legal oversight, investment in wastewater treatment, the introduction of desalination powered by SMRs, and the end of cartel corruption by severing Baja from the mainland power structures of narco-politics.

The United States has the resources. Baja has the need. What's missing is the political will to recognize this moment for what it is: a chance to build a bright future for both nations.

Chapter 10

Conclusion

Millions of years ago, the Colorado mountain range was lifted from an ancient sea to elevations above 16,000 feet. Along with the Sierra Nevada range, these mountains produce a rain shadow to the East and more precipitation along the coast to the West. The plate tectonics that produced these mountain ranges dealt with the Pacific plate subducting under the North American plate. Raising an ancient sea bed to a higher altitude created shale, oil, and Natural gas deposits that have yet to be fully developed or even explored. The heat and magnetic intrusions created massive regions of enriched mineral ores. The thought that Freedom Cities developed in the regions discussed could not support US citizens and their families with a sufficient income to acquire and maintain a household is laughable. The natural resources, environmental awareness, and technology are all in place.

US mining and petroleum production is the cleanest in the world. More thinning of these forests here is an environmental requirement. The real reason why the Arid Region has not previously been developed is, in the words of John Westly Powel, the downstream developers and politically powerful have monopolized the water rights for the region. Any precipitation that falls in most of these "Forsaken Zones" and collects into a stream or river no longer belongs to the land or population on which the precipitation fell. This paradigm extends to the point where it creates a political imbalance between the big coastal cities and more rural areas. Slowing population growth within the big cities and using "underutilized" portions of federal lands to create large numbers of small, well-spread, and economically connected Freedom Cities could be used to rebalance political power, reduce or solve the housing crisis and do so in a water-efficient environmentally sound manner. The regions circled in red below provide a starting point for the "site identification process.

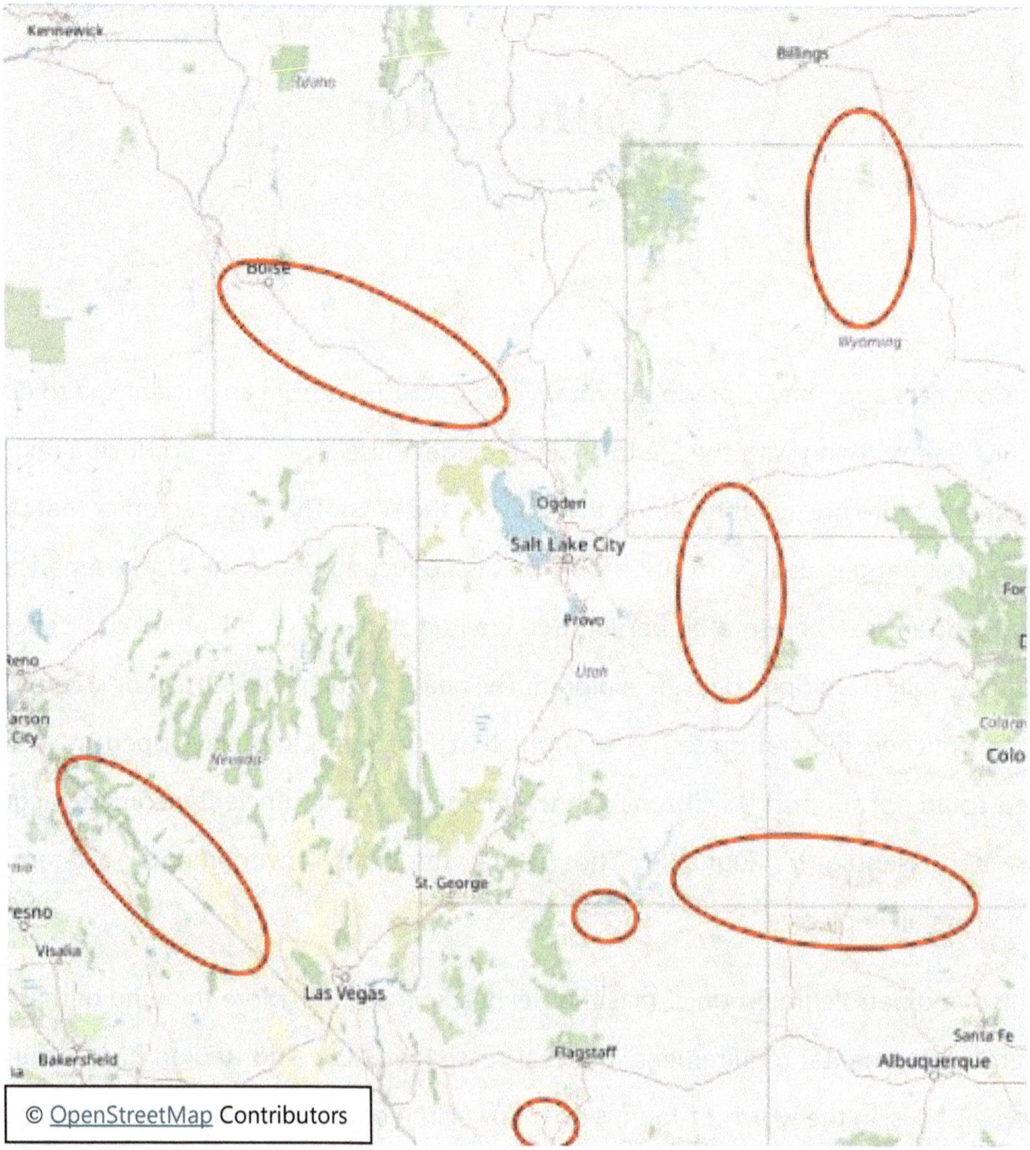

In all of these regions, a big yard will not be a requirement to install a high quality of life. Mother nature will provide your front and back yards. The school will have playgrounds and football and baseball fields created using artificial turf. If you have ever seen a college football game at Boise State University, the blue field gives away the fact that this can be done.

Indoor water use will be ample. You will not need to "run in circles" to get wet in the shower or flush the toilet more than one time. In today's world, and especially when building a new city from scratch, recycling/reclaiming the water we use indoors is an off-the-shelf technology. Restoring up to 95% of the water we use indoors back to nature via an advanced water reclamation system is a practice already in use in the Arid West.

Outdoor water use is where we will need to focus our conservation effort. Reclaiming evaporated or transpired water vapor is a far more difficult and expensive aspiration of life on planet Earth. Methods and opportunities do exist, but implementing an efficient version of such technologies remains an economic and environmental challenge for future generations.

Back East, a forward-looking modular builder, has built a 32-story tall housing and retail skyscraper that competes well with similar structures built using the more traditional "site" built approach. Modular skyscrapers like this use steel-framed mobile home-sized LEGO blocks that are lifted into position at the job site. Other builders use wood-framed LEGOS that limit the structure to four or five stories above the concrete foundation, parking structure, or commercial space. In either case, using the Henry Ford assembly line approach to manufactured housing reflects a shift in the housing market that has started to take hold. Building more factories that could take advantage of AI and robotics could provide a means of spreading small modular freedom cities (SMFCs) from the "forsaken zones" deep within the arid region to the less arid regions of the heartland and off of federal lands. AI and robotics are an inevitability that every market sector will need to face. In housing, the advantages include shifting to a higher tolerance factory-built solution that is faster and less expensive than hiring a workforce that has increasingly used illegal immigrants to build tracts of McMansions with questionable quality and social value to the younger generations that can't afford such housing.

President Trump has come up with yet another brilliant concept to resolve one of the inherent issues of our time. All citizens, taxpayers, and even investors should be hungry for more information about how the concept of freedom cities will work in practice. It will affect our lives and those of our children. The concepts in this book are meant to start a public conversation about what these cities will look like, where they will be built, and how they will function.

Please remember the resilience and productivity of the American People. The obstacles to building a better housing future for ourselves and our children are all about politics. We can work with the Trump administration and develop a better future, or we can leave it up to the bureaucrats. We have all seen how the latter works out.

References

All-American Canal | Imperial Irrigation District. (n.d.). Imperial Irrigation District. https://www.iid.com/water/water-transportation-system/colorado-river-facilities/all-american-canal

American Rivers. (2024, April 16). *Tijuana River Named Among America's Most Endangered Rivers® of 2024*. https://www.americanrivers.org/media-item/tijuana-river-named-among-americas-most-endangered-rivers-of-2024/

Assessment of Water Availability Issues Within the Legal Delta. (2021). In *Water Unavailability Methodology for the Delta Watershed*. https://www.waterboards.ca.gov/drought/drought_tools_methods/docs/wua_app_d_082021.pdf

California Aqueduct. (n.d.). Water Education Foundation. https://www.watereducation.org/aquapedia/california-aqueduct

California Governor Requests $40 Billion For Fire Aid, Washington Post Reports. (2025, February 22). Reuters. https://www.reuters.com/world/us/california-governor-requests-40-billion-fire-aid-washington-post-reports-2025-02-22/

Carlsbad Desalination Plant. (n.d.). Mott MacDonald. https://www.mottmac.com/en-us/projects/carlsbad-desalination-plant/

Carlsbad Desalination Plant Celebrates 5th Anniversary. (2020, December 14). Water News Network. https://www.waternewsnetwork.com/carlsbad-desalination-plant-celebrates-5th-anniversary/

Carlsbad Seawater Desalination Project Water Purchase Agreement. (2012, October 30). City of San Diego. https://www.sandiego.gov/sites/default/files/legacy/citycouncil/committees/wpitf/pdf/121030item3.pdf

Chappelle, C., & Hanak, E. (2021, May). *Water Affordability in California*. Public Policy Institute of California. https://www.ppic.org/publication/water-affordability/

City News Service. (2022, May 26). San Diego County Water Authority Proposes 5.2% Increase, Citing Inflation and Energy Costs. *Nbcsandiego*. https://www.nbcsandiego.com/news/local/san-diego-county-water-authority-proposes-5-2-increase-citing-inflation-and-energy-costs/2957211/

Claude "Bud" Lewis Carlsbad Desalination Plant. (n.d.). Kiewit Corporation. https://www.kiewit.com/projects/claude-bud-lewis-carlsbad-desalination-plant/

Colorado River Aqueduct. (n.d.). American Society of Civil Engineers. https://www.asce.org/about-civil-engineering/history-and-heritage/historic-landmarks/colorado-river-aqueduct

Colorado River Water: Good to the Last Acre-Foot. (n.d.). SIAM. https://www.siam.org/media/ljkpwaoe/m3challenge_colorado_river_water.pdf

Contra Costa Canal. (n.d.). Contra Costa Water District. https://www.ccwater.com/212/Contra-Costa-Canal

Contra Costa Water District. (2024). FY25–FY26 Budget. In *Contra Costa Water District*. https://ccwater.com/Archive/ViewFile/Item/396

Contra Costa Water District Water Services. (2008, April 9). Contra Costa LAFCo. https://www.contracostalafco.org/municipal_service_reviews/central_county_water_wastewater/4.0%20CCWD%20Water%20Wastewater%20Final.pdf

Cook, D. (2007). *Chino Valley Approves Pipeline*. Center for Biological Diversity. https://www.biologicaldiversity.org/news/media-archive/BigChino5-12-07.pdf

Delta Conveyance Project. (n.d.). In *Wikipedia*. Retrieved April 14, 2025, from https://en.wikipedia.org/wiki/Delta_Conveyance_Project

Demographics of Los Angeles County. (n.d.). Retrieved April 23, 2025, from https://en.wikipedia.org/wiki/Demographics_of_Los_Angeles_County

Desalination. (n.d.). Water Education Foundation. https://www.watereducation.org/aquapedia/desalination

Drinking Water Sources: An Overview. (2024, January 10). CDC. https://www.cdc.gov/drinking-water/about/drinking-water-sources-an-overview.html

Earthquake Facts: The Northridge Earthquake: January 17, 1994. (n.d.). Earthquake Country Alliance. https://www.earthquakecountry.org/northridge/facts/

Epps, R. (2025, January 31). *Navigating San Diego's Economic Future: Insights from the 41st Annual Economic Roundtable*. San Diego Workforce Partnership. https://workforce.org/news/navigating-san-diegos-economic-future-insights-from-the-41st-annual-economic-roundtable/

Forecasting California's Earthquakes—What Can We Expect in the Next 30 Years? (2007). In *USGC*. https://www.usgs.gov/index.php/publications/probability-one-or-more-m-7-earthquakes-southern-california-30-years

Friant-Kern Canal. (n.d.). In *Wikipedia*. Retrieved November 3, 2024, from https://en.wikipedia.org/wiki/Friant-Kern_Canal

Giddens, J. K. (2023, November 22). *Big Chino Pipeline still probably a decade out*. Sedona Red Rock News. https://www.redrocknews.com/2023/11/22/big-chino-pipeline-still-probably-a-decade-out/

Haney, A. J., Turner, S. D., & Springer, E. A. (2008). Ecological Implications of Verde River Flows. In *Arizona Conservation Science*. https://azconservation.org/dl/TNCAZ_VerdeRiver_Ecological_Flows.pdf

Hartig, T., & Jeyaram, S. (n.d.). *Reverse Osmosis in Desalination Plants*. University of Pittsburgh. https://sites.pitt.edu/~budny/papers/8117.pdf

Henry, L. (2022, April 26). *Third year of Drought Pushes Price of Water To $2,000 An Acre Foot*. SJV Water. https://sjvwater.org/third-year-of-drought-pushes-price-of-water-to-2000-an-acre-foot/

Hetch Hetchy. (n.d.). In *Wikipedia*. Retrieved February 16, 2025, from https://en.wikipedia.org/wiki/Hetch_Hetchy

Hetch Hetchy Water and Power System. (2024, August 14). Maven's Notebook. https://mavensnotebook.com/explainers/where-does-californias-water-come-from/hetch-hetchy-aqueduct/

Hoover, M., & Lucy, I. (2025, May 17). *The State of Housing in America*. U.S. Chamber of Commerce. https://www.uschamber.com/economy/the-state-of-housing-in-america

How Much Federal Money Goes Toward California State and Local Government? (n.d.). USA Facts. https://usafacts.org/answers/how-much-money-does-the-federal-government-provide-state-and-local-governments/state/california/

Hyperion Advanced Water Purification Facility (CA). (n.d.). Water Collaborative Delivery Association. https://watercollaborativedelivery.org/project/hyperion-advanced-water-purification-facility/

Infrastructure Investment and Jobs Act. (n.d.). Federal Aviation Administration. https://www.faa.gov/iija/airport-infrastructure

Ingram, J., & Sarwate, J. (2025, January 14). U.S. When L.A. Fires Broke Out, the 117-million Gallon Santa Ynez Reservoir Near Pacific Palisades Was Empty. Here's What We Know. *CBS News*. https://www.cbsnews.com/news/la-fires-santa-ynez-reservoir-pacific-palisades-california/

ISS: Astronauts Have Recycled 98% Of Water from Their Pee, Sweat and Breath. (2023, June 27). BBC. https://www.bbc.co.uk/newsround/66017072

January 2025 - Salton Sea Management Program. (n.d.). Salton Sea Management Program. https://saltonsea.ca.gov/2025/01/

Kern Water Bank Authority. (n.d.). Kern Water Bank Authority. https://www.kwb.org/

Laura, A. H., Vincent, & Hardy, C. (2020, February 21). *Federal Land Ownership: Overview and Data*. Congress.gov. https://www.congress.gov/crs-product/R42346

Lo, M., & Famiglietti, J. S. (2013). Irrigation in California's Central Valley Strengthens the Southwestern U.S. Water Cycle. *Geophysical Research Letters*, *40*(2), 301–306. https://doi.org/10.1002/grl.50108

Manhattan Population 2022. (n.d.). Azurewebsites. https://poeent.azurewebsites.net/manhattan

McAllister, R. (2024, September 25). *Boston Population 2025 – Key Statistics and Trends*. North American Community Hub Statistics. https://nchstats.com/boston-population/#google_vignette

Metropolitan to Further Reduce Reliance on Colorado River with $186 Million Federal Investment. (2025, January 10). The Metropolitan Water District of Southern California. https://www.mwdh2o.com/press-releases/metropolitan-to-further-reduce-reliance-on-colorado-river-with-186-million-federal-investment/

Mokelumne Aqueduct. (n.d.-a). Retrieved May 4, 2025, from https://en.wikipedia.org/wiki/Mokelumne_Aqueduct

Mokelumne Aqueduct. (n.d.-b). Maven's Notebook. https://mavensnotebook.com/explainers/where-does-californias-water-come-from/mokelumne-aqueduct/

Mullane, S. (2024, May 15). *There May Be a Wetter Future Ahead for the Colorado River, New CU Boulder Research Says*. Water Education Colorado. https://watereducationcolorado.org/fresh-water-news/colorado-river-drought-cycles-cu/

New Study Reveals Many American Families Struggle to Afford Basic Personal Care Items and Household Goods. (2013, September 19). Feeding America. https://www.feedingamerica.org/about-us/press-room/new-study-reveals-many-american-families-struggle-to-afford-basic-personal-care-items-and-household-goods

Ocean Water Properties. (n.d.). Hurricanes: Science and Society. https://www.hurricanescience.org/science/basic/water/index.html

Oil and Gas Facts & Figures 2024. (n.d.). Petroleum Association of Wyoming. https://pawyo.org/facts-figures/

Ostler, D. (Director). (2017, September 30). *Sixty-Ninth Annual Report of The Upper Colorado River Commission*. Upper Colorado River Commission. http://www.ucrcommission.com/RepDoc/UCRCAnnualReports/69_UCRC_Annual_Report.pdf

Owens Valley Water History (Chronology). (2008, January). County of Inyo California. https://www.inyowater.org/documents/reports/owens-valley-water-history-chronology/

Pros and Cons of High Recovery Reverse Osmosis Systems. (n.d.). Puretec Industrial Water. https://puretecwater.com/resources/pros-and-cons-of-high-recovery-reverse-osmosis-systems/

QuickFacts Los Angeles city, California. (n.d.). Census. https://www.census.gov/quickfacts/fact/table/losangelescitycalifornia/PST045224

Rance, H. (n.d.). 19 Crops That Are Grown Only in California. *CA Grown Blog*. https://californiagrown.org/blog/only-in-california/

Ring, E. (2025, April 30). *Ringside: How to Add 10 Million Acre Feet Per Year to California's Water Supply*. California Globe. https://californiaglobe.com/fr/ringside-how-to-add-10-million-acre-feet-per-year-to-californias-water-supply/

Salinity. (2025, March 11). Delta Stewardship Council. https://viewperformance.deltacouncil.ca.gov/pm/salinity

Salton Sea. (n.d.). Retrieved April 30, 2025, from https://en.wikipedia.org/wiki/Salton_Sea

Seawater Desalination: The Claude "Bud" Lewis Desalination Plant and Related Facilities. (2020, February). San Diego County Water Authority. https://www.sdcwa.org/sites/default/files/desal-carlsbad-fs-single.pdf

Seawater Desalination the Claude "Bud" Lewis Desalination Plant and Related Facilities. (2025, February). San Diego County Water Authority. https://www.sdcwa.org/wp-content/uploads/2020/11/desal-carlsbad-fs.pdf

Semancik, A. (2025, February 6). *The Economics of a Disaster: How the LA Wildfires May Impact the Economy*. OHIO University. https://www.ohio.edu/news/2025/02/economics-disaster-how-la-wildfires-may-impact-economy

Sharp, S. (2019, November 18). *First Glimpse of $54-Million Supportive Housing Project in DTLA*. Urbanize LA. https://la.urbanize.city/post/first-glimpse-54-million-supportive-housing-project-dtla

Smith, E. J. (2022, June 17). Poseidon Desal Plant in Carlsbad Needs $159M Upgrade. Ratepayers Will Foot the Bill. *The San Diego Union-Tribune*. https://www.sandiegouniontribune.com/2022/06/17/poseidon-desal-plant-in-carlsbad-needs-159m-upgrade-ratepayers-will-foot-the-bill/

Sorvino, C. (2021, September 20). *Amid Drought, Billionaires Control a Critical California Water Bank*. Forbes. https://www.forbes.com/sites/chloesorvino/2021/09/20/amid-drought-billionaires-control-a-critical-california-water-bank/

Spencer, M. R., Miniño, A. M., & Warner, M. (2022). Drug Overdose Deaths in the United States, 2001–2021. In *NCHS Data Brief* (No. 457). Centers for Disease Control and Prevention. https://www.cdc.gov/nchs/data/databriefs/db457.pdf

Staff, M. (2025, April 29). *Mexico Reaches Agreement to Send More Water to Southern US*. Mexico News Daily. https://mexiconewsdaily.com/news/mexico-us-reach-water-deliveries-agreement-texas-farmers-relief/

Statista. (2024, August 21). *Mexico: Population in Baja California 2008-2022*. https://www.statista.com/statistics/1038033/mexico-total-population-baja-california/

The Sacramento– San Joaquin Delta. (2016). In *Public Policy Institute of California*. https://www.ppic.org/wp-content/uploads/content/pubs/report/R_1016JM5R.pdf

Tourism to Grand Canyon National Park Contributed $710 Million To Local Economy in 2021. (2022, June 27). [Press release].

https://www.nps.gov/grca/learn/news/visitor-use-spending-to-grand-canyon-2021.htm

Tourism to Grand Canyon National Park Contributed $759 Million To Local Economy in 2022. (2023, August 21). [Press release]. https://www.nps.gov/grca/learn/news/2022-visitor-spending-to-grand-canyon-national-park.htm

United States Population Density. (n.d.). Kontur Inc. https://www.kontur.io/datasets/population-dataset/united-states-population-density/

U.S. Census Bureau QuickFacts: Wyoming. (n.d.). Census. https://www.census.gov/quickfacts/fact/table/WY/PST045224

U.S. Overdose Deaths In 2021 Increased Half as Much as in 2020 – But Are Still Up 15%. (2022, May 11). Centers for Disease Control and Prevention. https://www.cdc.gov/nchs/pressroom/nchs_press_releases/2022/202205.htm

U.S. Water News Online. (2001, April 20). *Nebraska Preserves Water Flows Under Agreement with Wyoming*. Wastewater Digest. https://www.wwdmag.com/home/news/10916547/nebraska-preserves-water-flows-under-agreement-with-wyoming

Water Purchase Agreement for Seawater Desalination Nearing Completion. (2012, August 9). San Diego County Water Authority. https://www.sdcwa.org/water-purchase-agreement-for-seawater-desalination-nearing-completion/

Water Systems. (2025). Tracking California. https://trackingcalifornia.org/topics/water-systems#gsc.tab=0

With Financing Complete, Poseidon Inks Deals for Construction, Operation of Carlsbad Desalination Project. (2012, December 24). Poseidon Water. http://www.poseidonwater.com/news-and-events/with-financing-complete-poseidon-inks-deals-for-construction-operation-of-carlsbad-desalination-project

www.ingramcontent.com/pod-product-compliance
Lightning Source LLC
LaVergne TN
LVHW082335200926
848481LV00012B/3352
* 9 7 9 8 3 4 9 4 4 6 5 7 3 *